Photocatalytic Semiconductors

Photocatalytic Semiconductors

Editor

Deepak Gupta

scitus
academics

Photocatalytic Semiconductors

Edited by **Deepak Gupta**

Printed in 2017

ISBN: 978-1-68117-226-2

Library of Congress Control Number: 2015936584

Contents

Preface

As early as 1969, a groundbreaking study informed to launch rigorous research activity on the reactivity of photoactivated semiconductor catalysts coined photocatalysis. Staring the decades of development, research activities have come a long way in accomplishments. Currently, articles in different aspects of photocatalysis abound most scientific journals and libraries. The frequent incidence of hazardous chemicals in wastewater produced by anthropogenic and industrial activities is of great concern because these pollutants contaminate lakes, rivers, and underground aquifers; furthermore, currently more pollutants, including traces of contaminants ranging from pharmaceutical drugs, hormones, and sunscreen to pesticides and dyes, are being detected at smaller concentrations in freshwater bodies. In addition, many of these contaminants are recalcitrant compounds, which cannot be degraded by the conventional methods of wastewater treatment; thus, many treated effluents that are considered "safe" for disposal still contain several toxic (bioactive) pollutants.

Editor

Chapter 1

Ag3PO4 Semiconductor Photocatalyst: Possibilities and Challenges

Department of Applied Physics, Key Laboratory for Micro-Nano Physics and Technology of Hunan Province, Hunan University, Changsha 410082, China

ABSTRACT

Ag_3PO_4 as a photocatalyst has attracted enormous attention in recent years due to its great potential in harvesting solar energy for environmental purification and fuel production. The photocatalytic performance of Ag_3PO_4 strongly depends on its morphology, exposed facets, and particle size. The effects of morphology and orientation of Ag_3PO_4 on the catalytic performance and the efforts on the stability improvement of Ag_3PO_4 are reviewed here. This paper also discusses the current theoretical understanding of photocatalytic mechanism of Ag_3PO_4, together with the recent progress towards developing Ag_3PO_4 composite photocatalysts. The crucial issues that should be addressed in future research activities are finally highlighted.

INTRODUCTION

The development of efficient photocatalysts is very important and desirable in environmental pollution mediation and solar energy conversion [1–6]. Over the past decades, fundamental progress has been made in developing novel photocatalysts, particularly visible light response catalysts for the efficient utilization of solar energy. A great deal of photocatalysts, including inorganic, molecular, and hybrid organic/inorganic materials, have been explored to meet specific requirements such as a light-absorbing wavelength modification, photoinduced charge separation, and a faster photocatalytic reaction. Among the various photocatalysts developed, TiO_2 is undoubtedly the most popular and widely used photocatalyst since it is of low cost, high photocatalytic activity, chemical and photochemical stability [7, 8]. However, TiO_2 is not ideal for all purposes and performs rather poorly in processes associated with solar photocatalysis due to its wide band gap (3–3.2 eV), thus making impractical overall technological process based on TiO_2. To design visible-light-driven photocatalysts, two strategies have been proposed. One is to modify the wide band gap photocatalysts (such as TiO_2, ZnS) by doping or by producing hetero-junctions between them and other materials [4, 7–12], and the other involves exploration of novel semiconductor materials capable of absorbing visible light. Various compounds, such as $BiVO_4$ [13], Bi_2WO_6 [1, 14], $CaBi_2O_4$ [15], $PbBi_2Nb_2O_9$ [16], $Bi_4Ti_3O_{12}$ [17], and Ag@AgCl [18], and others have been reported to be promising photocatalysts under visible light irradiation [19–22]. Despite many of these photocatalysts being effective for the degradation of organic pollutants and water splitting, up to date, the present achievements are still far from the ideal goal.

Yi and his coworkers [23] have recently presented the pioneering work on exploring the photocatalytic properties of Ag_3PO_4 that exhibit extremely high photooxidative capabilities for the O_2 evolution from water and the decomposition of organic dyes under visible-light irradiation. Actually, the photodegradation rate of organic dyes over Ag_3PO_4 is dozens of times faster than the rate over $BiVO_4$ and commercial $TiO_{2-x}N_x$ [23, 24]. Moreover, the most interesting is that this novel photocatalyst can achieve a quantum efficiency of up to 90% at wavelengths greater than 420 nm, which is significantly higher than the previous reported values. This finding potentially opens an avenue for solving current energy crisis and environment problems with abundant solar light, and the research of Ag_3PO_4 is thus attracting considerable interest. Since then, many efforts have been devoted to further improving and optimizing their photoelectric and photocatalytic properties. Despite the fact that Ag_3PO_4 is a promising candidate for environmental remediation and renewable energy, the consumption of a large amount of noble metal and the low structural stability of pure Ag_3PO_4 strongly limits its practical environmental applications. Therefore, it is a highly crucial task to improve the photocatalytic stability of Ag_3PO_4 while maintaining its high photocatalytic activity. In this paper, the effects of morphology and orientation of Ag_3PO_4 on the catalytic performance and the current theoretical understanding of key aspects of Ag_3PO_4 photocatalysts are presented. Also reviewed is the effort on the photocatalytic stability of Ag_3PO_4 and Ag_3PO_4 composite photocatalyst.

THE STRUCTURE OF AG$_3$PO$_4$

Ag_3PO_4 is of body-centred cubic structure type with space group P4-3n and a lattice parameter of ~6:004 Å. The structure consists of isolated, regular PO_4 tetrahedra (P–O distance of ~1.539 Å) forming a body-centred-cubic lattice. The six Ag^+ ions are distributed among twelve sites of twofold symmetry [25]. This indicates that each Ag atom at (0.25, 0, 0.50) actually occupies one of the two sites at $(x, 0, 0.50)$ and $(0.5 - x, 0, 0.50)$ on the 2-fold axis. The unit-cell structure of cubic Ag_3PO_4 is shown in Figure 1, in which the Ag atom experiences 4-fold coordination by four O atoms [24]. The P atoms have 4-fold coordination surrounded by four O atoms, while the O atoms have 4-fold coordination surrounded by three Ag atoms and one P atom.

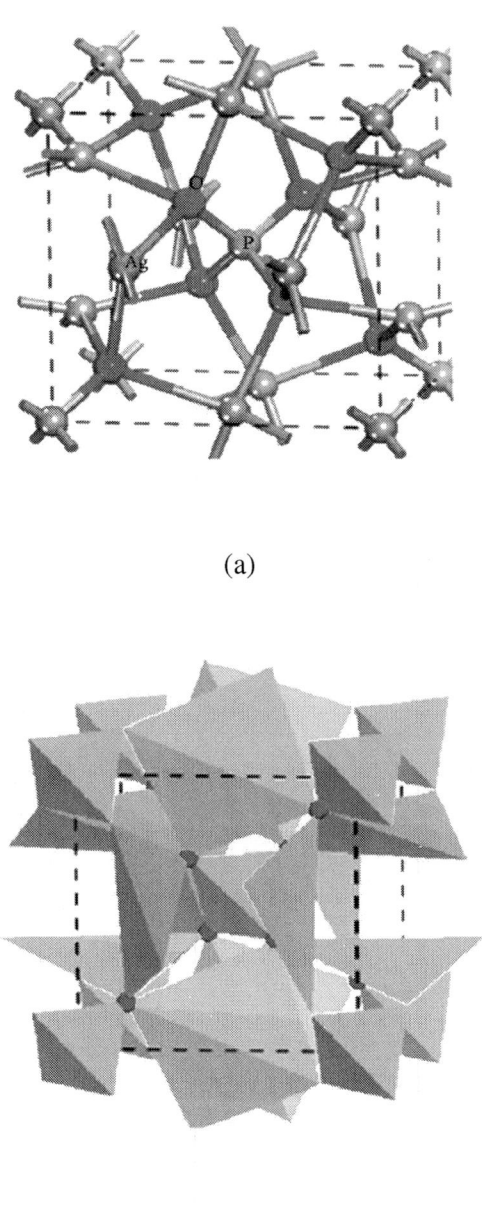

(a)

(b)

Figure 1: Unit-cell structure of cubic Ag_3PO_4, showing (a) ball and stick and (b) polyhedron configurations. Red, purple, and blue spheres represent O, P, and Ag atoms, respectively [24].

MORPHOLOGY CONTROL AND CATALYTIC PROPERTIES OF AG$_3$PO$_4$

Since Ag$_3$PO$_4$ was first reported as visible light response photocatalyst by Yi and his coworkers [23], much research has been devoted to investigate this active photocatalyst [26–29], and different methods have been developed to synthesize various Ag$_3$PO$_4$ and its composites [28, 30–32]. The investigation of photocatalytic activity shows that all of the prepared Ag$_3$PO$_4$ exhibit excellent photocatalytic activity under UV light irradiation or visible-light illumination, which is much more excellent than that of commercial P25 TiO$_2$ [33,34] or N-doped TiO$_2$ [30, 35, 36] as shown in Figure 2.

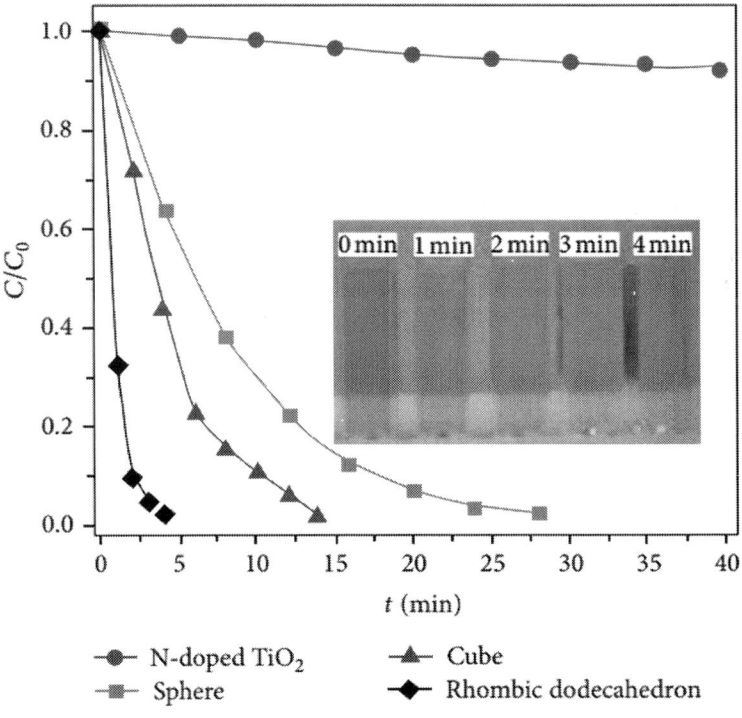

(a)

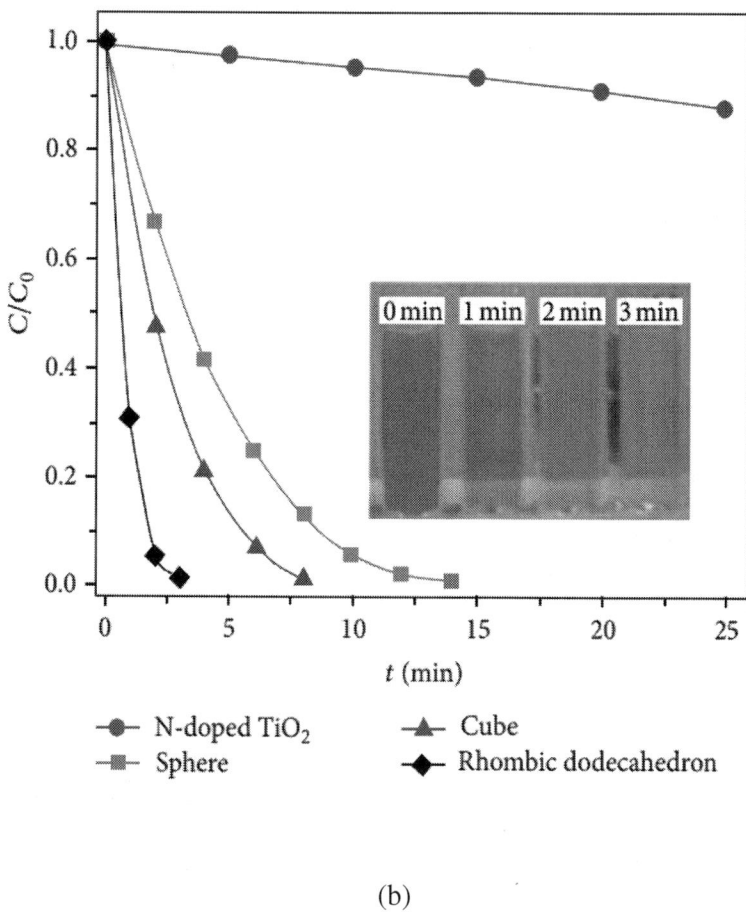

(b)

Figure 2: Photocatalytic activities of Ag_3PO_4 rhombic dodecahedrons, cubes, and spheres and N-doped TiO_2 for (a) MO and (b) RhB degradation under visible-light irradiation ($\lambda > 420$ nm) [30].

It is well known that the morphology of materials is closely related to the exposed facets of the crystals, which directly affect the properties of the catalysts. Various Ag_3PO_4 nanostructures including spherical morphology [33, 34], rhombic dodecahedrons [30], concave trisoctahedrons [37], cubes [26, 30], and tetrapods [38] with controlling particle size [39] have been designed and synthesized to further improve or optimize the photocatalytic properties. The investigations revealed that the rhombic dodecahedrons exhibit much higher photocatalytic activity in comparison with spheres and cubes for the degradation of organic contaminants [23,30], and the enhanced photocatalytic activity of Ag_3PO_4 rhombic dodecahedrons is primarily ascribed to the higher surface energy of {110} facets (1.31 J/m²). The highly exposed {110} facets also lead to the higher visible light activity of Ag_3PO_4 tetrapods than that of the polyhedrons for the degradation of toxic organic compounds [38]. The concave trisoctahedral Ag_3PO_4 microcrystals with high-index facets have also been reported to show much higher photocatalytic properties than cubic Ag_3PO_4 and commercial N-doped TiO_2 [37].

The synthetic parameters, such as reaction component, temperature, and reaction time, usually serve as effective routes for tailoring the morphology and structure of Ag_3PO_4 crystals. By directly reacting $AgNO_3$ and Na_2HPO_4, only Ag_3PO_4 particles with irregular spherical structures can be obtained, while the single crystalline Ag_3PO_4 submicrocubes with sharp corners, edges, and smooth surfaces can be fabricated with $[Ag(NH_3)_2]^+$ as precursors under the same conditions, which therefore exhibit higher photocatalytic activity than spherical nanoparticles [35]. Moreover, the morphologies and structures of the as-prepared Ag_3PO_4 crystals could be further tailored by changing the aging time of the $[Ag(NH_3)_2]^+$ complex due to the ammonia volatility. For example, when $[Ag(NH_3)_2]^+$ complex solution was aged at room temperature for 1 h, the as-prepared Ag_3PO_4 crystals possess irregular cubic morphology and six rhombic planes. Upon further increasing the aging time up to 2 h, the Ag_3PO_4 crystals transform into irregular structure with concave surfaces. Interestingly, when the aging time has been prolonged to 4 h, the corresponding Ag_3PO_4 products with four symmetrically arrowheaded morphologies have been fabricated (shown in Figure 3), and no cubes have been observed. The variation of morphologies and structures of the as-prepared Ag_3PO_4 crystals leads to the variation of optical properties (Figure 3(d)). Similar morphology variation of Ag_3PO_4 particles prepared by a pyridine-assisted hydrothermal method is found to change regularly from spherical to particles with the corners and edges then to rhombic dodecahedrons with an increase in pyridine concentration [27].

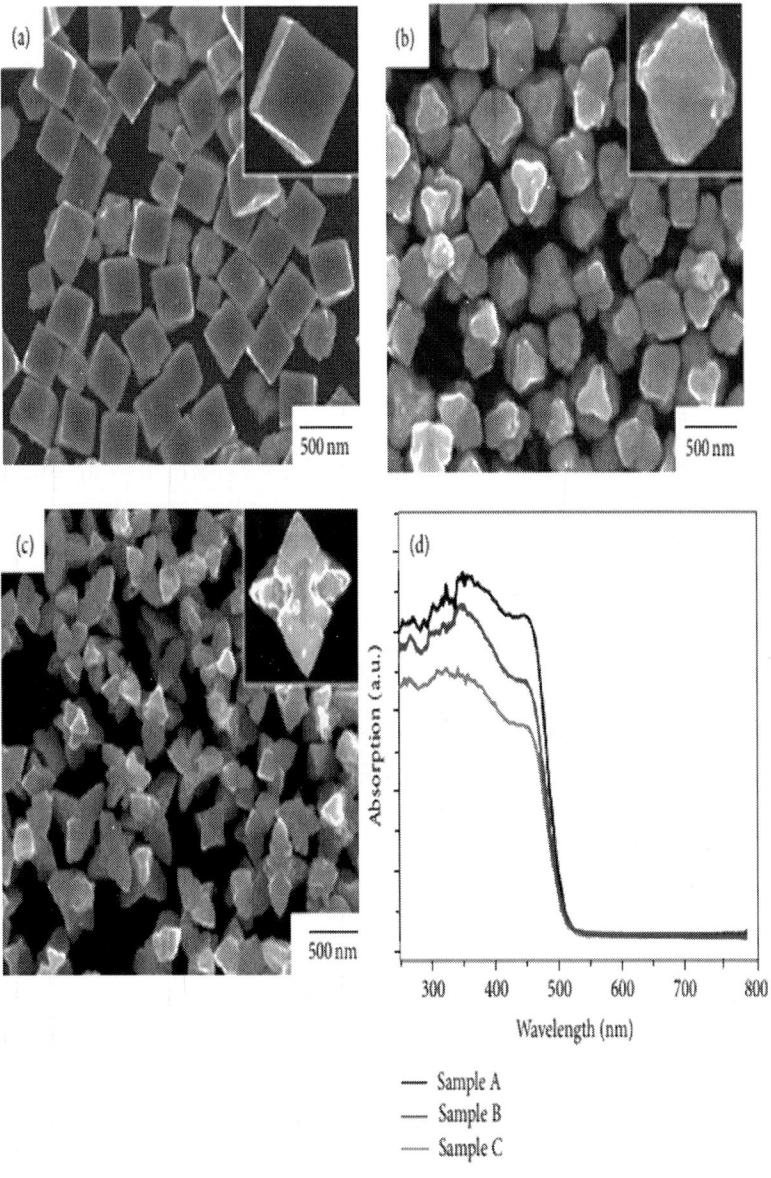

Figure 3: SEM images of Ag_3PO_4 products prepared with different aging time: (a) 1 h, (b) 2 h, and (c) 4 h; (d) their ultraviolet-visible diffusive absorption spectrums [35].

THEORETICAL STUDY OF THE ELECTRONIC STRUCTURE AND PHOTOCATALYTIC PERFORMANCE

To elucidate its mechanism of the extremely high photooxidative under visible light irradiation of Ag_3PO_4, theoretical works have been carried out by using first-principle method. Photocatalysis reactions are determined primarily by three processes: (1) the photoexcitation of electron-hole pairs due to light harvesting, (2) the transfer of carriers to the surface, and (3) chemical reactions on the surface. So far, theoretical investigations are mainly focused on the first process, which is relevant to energy band configuration because photoexcited carriers are generated only when the incident photon energy is higher than the bandgap of the photocatalyst. Alignment between the band edges and the redox potentials of the target molecules is also crucial, considering that photoexcited carriers can only be transferred to the adsorbed molecules when there is a sufficiently large negative offset of the conduction band minimum (CBM) and a sufficiently large positive offset of the valence band maximum (VBM) with respect to the redox potentials. Ma et al. investigated the electronic properties and photocatalytic activation of Ag_3PO_4 using first principles density functional theory (DFT) incorporating the local density approximation (LDA) + U formalism [24]. It is found that one PO_4 tetrahedron and three tetrahedral AgO_4 are combined with each other through the corner oxygen. The tetrahedral AgO_4 is heavily distorted to have a dipole moment of 2.2 D (D = debye), which should be closely attributed to a specific nonmetal salt of the oxyacid structure of Ag_3PO_4. In addition,

PO_4^{3-} ions have a large negative charge which maintains a large dipole in the Ag_3PO_4, resulting in the distortion of tetrahedral AgO_4. As a consequence, a correlation between the photocatalytic activity and the distortion of AgO_4 tetrahedron is true for a specific phosphate structure.

Meanwhile, PO_4^{3-} possessing a large electron cloud overlapping prefers to attract holes and repel electrons, which helps the e⁻/h⁺ separation. The calculated results show that Ag_3PO_4 is an indirect band gap semiconductor; the direct band gap is 2.61 eV and the indirect band gap is 2.43 eV having the character of or ϖ bonding states in the VB and the corresponding * or ϖ* antibonding states in the CB. The effective mass of electrons is far smaller than that of holes, which consequently results in a striking difference of the mobility between photoexcited electrons and holes. This is beneficial for decreasing the recombination of electron-hole pairs in Ag_3PO_4. In addition, Ag vacancies in Ag_3PO_4 with high concentration have a significant effect on the separation of electron-hole pairs and optical absorbance in the visible-light region. To explore the mechanism of the high performance of Ag_3PO_4, Umezawa et al. have conducted a comparative study of the electronic structures of Ag_3PO_4, Ag_2O, and $AgNbO_3$ by first principles calculations [32]. The CBM consisting of Ag s states in Ag_3PO_4 and Ag d states in Ag_2O is the main difference between Ag_3PO_4 and Ag_2O, which is due to the rigid PO_4 tetrahedral unit; partial charge density originating from O sp and P sp derived bonding states is evident in the local density of states (DOS). This increases the ionic character of Ag^+ and $(PO_4)^{3-}$, making the covalent Ag–O bonds weaker. Thus, the hybridization of Ag d and O p states is negligible in Ag_3PO_4, rendering fully occupied d states at the VBM. On the contrary, covalent Ag–O bonds, that is, antibonding states of Ag d–O p lying at the CBM are formed in Ag_2O. In $AgNbO_3$, the majority of the CBM is composed of Nb d states and the nature of the Ag–O bonding does not significantly affect the characteristics of the CBM.

In Figure 4, the band structures of Ag_3PO_4, Ag_2O, and $AgNbO_3$ are illustrated, accompanied by the wave function corresponding to the CBM. In Ag_3PO_4, the wave function at the CBM is dramatically delocalized due to mainly consisting of Ag s states. In contrast, the CBM of Ag_2O possesses the character of Ag d–O p antibonding states, which are unfavorable for electron transfer due to their localization. In Ag_3PO_4, a large amount of hybridization between Ag s states on adjacent atoms leads to a dispersive band structure at the CBM without the "contamination" of d states, resulting in a decreased effective electron mass. In $AgNbO_3$, the CBM mainly composes of d_{xz} and d_{yz} states, causing an anisotropic charge distribution (Figure 4(c)) and allowing electron transfer only along the z-direction. This anisotropy can also be found in the band structure, where the dispersion of the CBM is very dispersive along X-S-Y and flat along S- indicating the high dependence of the directions. This is undesirable for photocatalysis process because the limited directionality of the electron transfer leads to higher probability of carrier recombination. The CBM of Ag_3PO_4 has, on the contrary, a very isotropic distribution and is hence favorable for electron transfer

(Figure 4(a)). The effective mass of the electron is relatively small in every direction in Ag_3PO_4, which is also advantageous for electron transfer. Thus, the excellent photocatalytic performance of Ag_3PO_4 is partly due to the highly dispersive band structure at the CBM, which results from Ag s–Ag s hybridization without localized d states.

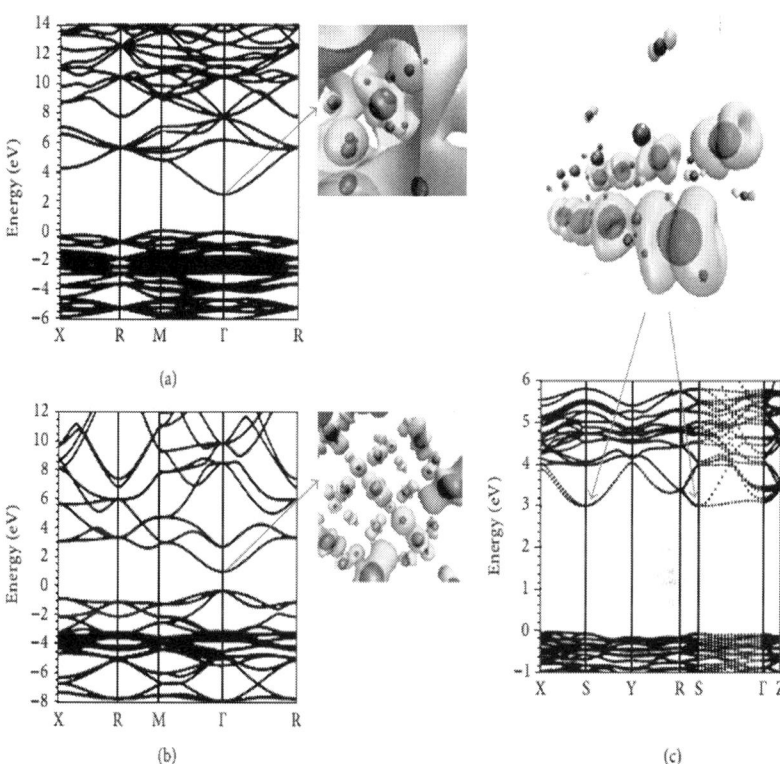

(a)

(b)

(c)

Figure 4: (Color) band structures for (a) Ag_3PO_4, (b)Ag_2O, and (c) $AgNbO_3$. The square of the wave function (yellow surface) corresponding to the CBM is also shown in each case, where silver, red, mauve, and green particles represent the positions of Ag, O, P, and Nb atoms, respectively. The isosurfaces are at 0.01 e/Å³ [32].

Using the local density approximation (LDA) and LDA+U approaches, the calculated band gaps of 0.36 and 1.30 eV are far less than the experimental value of 2.45 eV due to the missing discontinuity in the exchange-correlation potential and the self-interaction error within the LDA. To effectively remedy the drawbacks of LDA and understand the photocatalytic mechanism, hybrid density functional theory (DFT) using PBE0 formalism, was used to calculate the band structure, density of state (DOS), and optical properties of Ag_3PO_4 [40]. The hybrid-DFT method gives a direct band gap, $E_{gdir}=2.61eV$, and an indirect one, $E_{gindir}=2.43eV$, which agrees well with the experimental value (2.45 eV). Compared with the top of valence band (VB), the bottom of conduction band (CB) is well dispersive, which indicates that the photogenerated electrons possess smaller effective mass and, therefore, higher migration ability. The redox ability of Ag_3PO_4 is evaluated by determining the energy positions of valence and conduction bands using Mulliken electronegativity and the band gap value calculated accurately. Figure 5 renders the valence and conduction band edge potentials of Ag_3PO_4. The VBM potential of Ag_3PO_4 is 2.67 V, more positive than O_2/H_2O (1.23 V), indicating that Ag_3PO_4 has the ability to oxidize H_2O to produce O_2 or oxidation pollutants. Whereas the CBM potential of Ag_3PO_4 is 0.24 V, which is lesser than H^+/H_2 (0 V) and cannot reduce H^+ to H_2 [40]. More recently, Ma et al. investigated the electronic and photocatalytic properties of $Ag_3PC_4^{VI}$ (C = O, S, Se) by the hybrid density functional method. Similar results for Ag_3PO_4 are obtained [41].

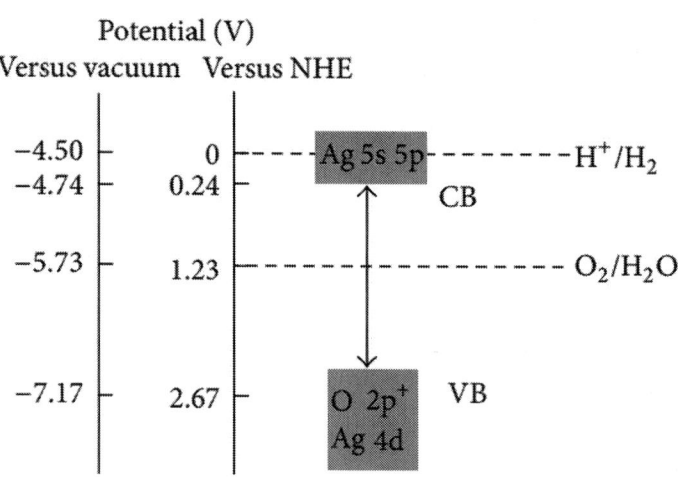

Figure 5: Calculated VBM and CBM potentials of Ag_3PO_4 [40].

VISIBLE LIGHT RESPONSE AG$_3$PO$_4$ BASED COMPOSITE PHOTOCATALYST

To harvest photons in visible region, many narrow bandgap metal oxides or chalcogenides have been coupled with TiO$_2$ to fabricate visible-light photocatalysts, which exhibit visible-light photocatalytic activity to a certain extent. Such a strategy is also applied to modify Ag$_3$PO$_4$ photocatalyst to enhance its photocatalytic activity and/ or improve its stability. The previous investigations showed that the photocatalytic activity of Ag$_3$PO$_4$ can be enhanced as Ag nanoparticles deposited on Ag$_3$PO$_4$ because the Ag$_3$PO$_4$ decomposition could capture the photogenerated electrons and thus prevent the recombination of electron-hole pairs within the Ag$_3$PO$_4$ samples at the initial stage of photocatalytic reactions. However, the photoactivity decreases with increasing Ag contents due to the formation of Ag layers on the surface of Ag$_3$PO$_4$ that shield light absorption, inhibit the transfer of holes from the valance band of Ag$_3$PO$_4$ to the interface between photocatalyst, and solution and also hinder the contact of dye molecules with Ag$_3$PO$_4$, and, accordingly, the photocatalytic activity deteriorates gradually [34]. This deterioration of the Ag$_3$PO$_4$ photocatalytic activity due to photocorrosion largely limits its practical application as a recyclable highly efficient photocatalyst. It is found that the Ag/Ag$_3$PO$_4$ heterocubes synthesized by reacting Ag$_3$PO$_4$ cubes with glucose in an aqueous ammonia solution exhibit higher photocatalytic activities than pure Ag$_3$PO$_4$ cubes for the organic contaminants degradation under visible-light irradiation [26]. The stability improvement of Ag$_3$PO$_4$ by covering Ag^0 nanoparticles on the surface of Ag$_3$PO$_4$ is attributed to the localized surface plasmon resonance (LSPR) effects of silver nanoparticles and a large negative charge of PO$_4^{3-}$ ions [2, 27], which effectively inhibit the reducibility of Ag$^+$ ions in the Ag$_3$PO$_4$ lattice. Ag$_3$PO$_4$ can also be rejuvenated from weak photocatalytically active Ag as a recyclable highly efficient photocatalyst by oxidizing Ag with H$_2$O$_2$ under a PO$_4^{3-}$ ion atmosphere [29]. However, these strategies are not ideal from the practical application perspective. Thus, the fabrication of Ag$_3$PO$_4$ based composite photocatalysts with high photocatalytic activity and excellent stability as well as lower Ag usage for their large scale applications is desirable.

Yao et al. synthesized Ag_3PO_4/TiO_2 visible light photocatalyst by depositing of Ag_3PO_4 nanoparticles onto the TiO_2 (P25) surface photocatalyst [42]. Their results show that the Ag_3PO_4/TiO_2 heterostructured photocatalyst shows enhanced activity and is much more stable than unsupported Ag_3PO_4. The enhanced activity is attributed to the electron-hole effective separation and the larger surface area of the Ag_3PO_4/TiO_2 composite, while the enhanced stability is ascribed to the chemical adsorption of Ag^+ cations in Ag_3PO_4 and O^- anions in TiO_2. Moreover, the silver weight percentage of the photocatalyst decreases from 77% to 47%, significantly reducing the cost of Ag_3PO_4 based photocatalysts for the Ag_3PO_4/TiO_2 composite [42].

The UV photocatalytic activity of Ag_3PO_4/TiO_2 composite heterostructures was comparable to that of Ag_3PO_4 nanoparticles surfaces. While the stability and hence reusability of the Ag_3PO_4/TiO_2 heterostructure catalysts was substantially enhanced as compared with that of Ag_3PO_4 nanoparticles or TiO_2 nanobelts alone. These results were attributed to the improved charge separation of the photogenerated electrons and holes under UV light at the Ag_3PO_4/TiO_2 interface and/or surfactant-like function of the nanobelts in stabilizing the Ag_3PO_4 nanoparticles. Ag_3PO_4/TiO_2 composite heterostructures appear to be more desirable in long-term applications because of their photocatalytic activity as well as the enhanced chemical stability [33].

Considering that the VB level of Ag_3PO_4 is appreciably lower than that of TiO_2 with +2.7 V (versus NHE) and Ag_3PO_4 can be severed as an appropriate sensitizer for TiO_2, Lee et al. fabricated the novel heterojunction structures of Ag_3PO_4-core/TiO_2-shell by covering the Ag_3PO_4 nanoparticles with polycrystalline TiO_2 by sol-gel method. The prepared Ag_3PO_4/TiO_2 composites show notably enhanced photocatalytic activity in decomposing gaseous 2-propanol and evolving CO_2 compared to bare Ag_3PO_4 and TiO_2. It is inferred that the unusually high visible-light photocatalytic activity of Ag_3PO_4/TiO_2 composite originates from the unique relative band positions of the two semiconductors [43].

Besides Ag_3PO_4/TiO_2 composite heterostructures, AgX/Ag_3PO_4 (X=Cl, Br, I) heterocrystals have also attracted much attention due to the excellent photocatalytic activity. Bi and coworkers [44] have reported that the AgX/Ag_3PO_4 (X=Cl, Br, I) heterocrystals prepared by in situ ionexchange method embodied some advantages compared to the single Ag_3PO_4, and it is a more promising and fascinating visible-light-driven photocatalyst than pure Ag_3PO_4 [45]. The $AgBr/Ag_3PO_4$ hybrid synthesized using an in situ anion-exchange method displayed much higher photocatalytic activity than single AgBr or Ag_3PO_4, as well as high stability under visible light irradiation. The high stability was attributed to the formed $Ag@AgBr/Ag_3PO_4@Ag$ plasmonic system, which effectively retains its activity due to the efficient transfer of photoinduced electrons [45].

CONCLUSION AND PERSPECTIVES

In this paper, we have summarized the survey of efforts on Ag_3PO_4 photocatalysts. As a quintessence, for example, Ag_3PO_4 photocatalysts have excellent photocatalytic activity and yield a high quantum yield of nearly 90% under visible light (λ =420nm) found for the evolution of O_2 in water photolysis [23]. All of the developments show that Ag_3PO_4 and/or its heterogeneous composites have the potential as excellent visible-light-active candidates with high photocatalytic activity. However, to comply with the challenging requirements of economically viable industrial production, many issues are still yet to be addressed, including long-term stability of Ag_3PO_4 in photocatalytic processes. New methods for fabrication Ag_3PO_4 with exposed high-energy facets and novel heterogeneous Ag_3PO_4 cocatalysts are highly desirable. To date, there have been relatively few quantitative studies in the charge carrier dynamics in Ag_3PO_4 and/or its cocatalysts systems, how these dynamics are related to material design, and how they impact the photocatalytic processes. In heterogeneous systems, a particular challenge is that electron-hole recombination is a bimolecular process, which its dynamics often depend nonlinearly on the charge carrier density and so cannot be described using a single time constant. Therefore, obtaining a substantial breakthrough in efficiency requires an exact understanding of the surface/interface processes at the atomic scale. A true picture of photocatalyst surfaces in action, including such recombination dynamics, needs to be studied by in situ observations using transient optical spectroscopy techniques. A further challenge is translating laboratory-scale academic research into scalable, manufacturable technologies to meet the demands of the efficient utilization of solar energy in the areas of renewable energy and environmental purification.

ACKNOWLEDGMENTS

This work is supported by the Hunan Provincial Natural Science Foundation of China (Grant no. 12JJ3009), the Science and Technology Plan Projects of Hunan Province (2012ZK3021).

REFERENCES

1. L. W. Zhang, Y. Man, and Y. F. Zhu, "Effects of Mo replacement on the structure and visible-light-induced photocatalytic performances of Bi_2WO_6 photocatalyst," ACS Catalysis, vol. 1, no. 8, pp. 841–848, 2011. · ·

2. Y. P. Liu, L. Fang, H. D. Lu, L. J. Liu, H. Wang, and C. Z. Hu, "Highly efficient and stable Ag/Ag_3PO_4 plasmonic photocatalyst in visible light," Catalysis Communications, vol. 17, pp. 200–204, 2012. ·

3. R. Asahi, T. Morikawa, T. Ohwaki, et al., "Photocatalysts sensitive to visible light," Science, vol. 295, no. 5555, pp. 626–627, 2002. · ·

4. G. S. Wu, J. L. Wen, S. Nigro, and A. C. Chen, "One-step synthesis of N-and F-codoped mesoporous TiO_2 photocatalysts with high visible light activity," Nanotechnology, vol. 21, no. 8, Article ID 085701, 2010. · ·

5. Y. Tian, G. F. Huang, L. J. Tang, M. G. Xia, W. Q. Huang, and Z. L. Ma, "Size-controllable synthesis and enhanced photocatalytic activity of porous ZnS nanospheres," Materials Letters, vol. 83, pp. 104–107, 2012. ·

6. Y. Chen, G. F. Huang, W. Q. Huang, B. S. Zou, and A. L. Pan, "Enhanced visible-light photoactivity of La-doped ZnS thin films," Applied Physics A, vol. 108, no. 4, pp. 895–900, 2012. ·

7. V. Etacheri, M. K. Seery, S. J. Hinder, and S. C. Pillai, "Highly visible light active $TiO_{2-x}N_x$ heterojunction photocatalysts," Chemistry of Materials, vol. 22, no. 13, pp. 3843–3853, 2010. · ·

8.   L. Feng, K. F. He, and W. P. Chen, "Study on stability of AgI/TiO$_2$ visible light photocatalysts in solutions of various pH values," New Materials, Applications and Processes, vol. 399-401, pp. 1272–1275, 2012.

9.   J. J. Yuan, H. D. Li, S. Y. Gao, Y. H. Lin, and H. Y. Li, "A facile route to n-type TiO$_2$-nanotube/p-type boron-doped-diamond heterojunction for highly efficient photocatalysts," Chemical Communications, vol. 46, no. 18, pp. 3119–3121, 2010.

10.   J. Q. Li, D. F. Wang, Z. Y. Guo, and Z. F. Zhu, "Preparation, characterization and visible-light-driven photocatalytic activity of Fe-incorporated TiO$_2$ microspheres photocatalysts," Applied Surface Science, vol. 263, pp. 382–388, 2012. ·

11.   Y. Komai, K. Okitsu, R. Nishimura et al., "Visible light response of nitrogen and sulfur co-doped TiO$_2$photocatalysts fabricated by anodic oxidation," Catalysis Today, vol. 164, no. 1, pp. 399–403, 2011. · ·

12.   X. M. Fang, Z. G. Zhang, and Q. L. Chen, "Nitrogen doped TiO$_2$ photocatalysts with visible-light activity," Progress in Chemistry, vol. 19, no. 9, pp. 1282–1290, 2007.

13.   Y. Hu, D. Z. Li, Y. Zheng et al., "BiVO$_4$/TiO$_2$ nanocrystalline heterostructure: a wide spectrum responsive photocatalyst towards the highly efficient decomposition of gaseous benzene," Applied Catalysis B, vol. 104, no. 1-2, pp. 30–36, 2011. · ·

14.   Y. J. Chen, Y. Q. Zhang, C. Liu, A. M. Lu, and W. H. Zhang, "Photodegradation of malachite green by nanostructured Bi$_2$WO$_6$ visible light-induced photocatalyst," International Journal of Photoenergy, vol. 2012, Article ID 510158, 6 pages, 2012. ·

15.   J. W. Tang, Z. G. Zou, and J. H. Ye, "Efficient photocatalytic decomposition of organic contaminants over CaBi$_2$O$_4$ under visible-light irradiation," Angewandte Chemie, vol. 43, no. 34, pp. 4463–4466, 2004. · ·

16.   H. G. Kim, P. H. Borse, J. S. Jang, E. D. Jeong, and J. S. Lee, "Enhanced photochemical properties of electron rich W-doped PbBi$_2$Nb$_2$O$_9$ layered perovskite material under visible-light irradiation,"Materials Letters, vol. 62, no. 8-9, pp. 1427–1430, 2008. · ·

17.   T. P. Cao, Y. J. Li, C. H. Wang et al., "$Bi_4Ti_3O_{12}$ nanosheets/ TiO_2 submicron fibers heterostructures: in situ fabrication and high visible light photocatalytic activity," Journal of Materials Chemistry, vol. 21, no. 19, pp. 6922–6927, 2011. · ·

18.   D. L. Chen, S. H. Yoo, Q. S. Huang, G. Ali, and S. O. Cho, "Sonochemical synthesis of Ag/AgCl nanocubes and their efficient visible-light-driven photocatalytic performance," Chemistry A, vol. 18, no. 17, pp. 5192–5200, 2012. ·

19.   M. Kitano, M. Matsuoka, M. Ueshima, and M. Anpo, "Recent developments in titanium oxide-based photocatalysts," Applied Catalysis A, vol. 325, no. 1, pp. 1–14, 2007. · ·

20.   L. Xu, C. Li, W. Shi, J. Guan, and Z. Sun, "Visible light-response $NaTa_{1-x}Cu_xO_3$ photocatalysts for hydrogen production from methanol aqueous solution," Journal of Molecular Catalysis A, vol. 360, pp. 42–47, 2012. ·

21.   R. Niishiro, H. Kato, and A. Kudo, "Nickel and either tantalum or niobium-codoped TiO_2 and $SrTiO_3$ photocatalysts with visible-light response for H_2 or O_2 evolution from aqueous solutions," Physical Chemistry Chemical Physics, vol. 7, no. 10, pp. 2241–2245, 2005. · ·

22.   W. Xiong, Q. D. Zhao, X. Y. Li, and D. K. Zhang, "One-step synthesis of flower-like Ag/AgCl/BiOCl composite with enhanced visible-light photocatalytic activity," Catalysis Communications, vol. 16, no. 1, pp. 229–233, 2011. ·

23.   Z. G. Yi, J. H. Ye, N. Kikugawa et al., "An orthophosphate semiconductor with photooxidation properties under visible-light irradiation," Nature Materials, vol. 9, no. 7, pp. 559–564, 2010. · ·

24.   X. G. Ma, B. Lu, D. Li, R. Shi, C. S. Pan, and Y. F. Zhu, "Origin of photocatalytic activation of silver orthophosphate from first-principles," Journal of Physical Chemistry C, vol. 115, no. 11, pp. 4680–4687, 2011. · ·

25.   H. N. Ng, C. Calvo, and R. Faggiani, "A new investigation of the structure of silver orthophosphate," Acta Crystallographica B, vol. 34, no. 3, pp. 898–899, 1978. ·

26.   Y. P. Bi, H. Y. Hu, S. X. Ouyang, Z. B. Jiao, G. X. Lu, and J. H. Ye, "Selective growth of metallic Ag nanocrystals on Ag_3PO_4

submicro-cubes for photocatalytic applications," Chemistry A, vol. 18, no. 45, pp. 14272–14275, 2012. ·

27. Y. P. Liu, L. Fang, H. D. Lu, Y. W. Li, C. Z. Hu, and H. G. Yu, "One-pot pyridine-assisted synthesis of visible-light-driven photocatalyst Ag/Ag$_3$PO$_4$," Applied Catalysis B, vol. 115-116, pp. 245–252, 2012. ·

28. G. P. Li and L. Q. Mao, "Magnetically separable Fe$_3$O$_4$-Ag$_3$PO$_4$ sub-micrometre composite: facile synthesis, high visible light-driven photocatalytic efficiency, and good recyclability," RSC Advances, vol. 2, no. 12, pp. 5108–5111, 2012. ·

29. H. Wang, Y. S. Bai, J. T. Yang, X. F. Lang, J. H. Li, and L. Guo, "A facile way to rejuvenate Ag$_3$PO$_4$ as a recyclable highly efficient photocatalyst," Chemistry A, vol. 18, no. 18, pp. 5524–5529, 2012. ·

30. Y. Bi, S. Ouyang, N. Umezawa, J. Cao, and J. Ye, "Facet effect of single-crystalline Ag$_3$PO$_4$ sub-microcrystals on photocatalytic properties," Journal of the American Chemical Society, vol. 133, no. 17, pp. 6490–6492, 2011. · ·

31. H. C. Zhang, H. Huang, H. Ming et al., "Carbon quantum dots/Ag$_3$PO$_4$ complex photocatalysts with enhanced photocatalytic activity and stability under visible light," Journal of Materials Chemistry, vol. 22, no. 21, pp. 10501–10506, 2012. ·

32. N. Umezawa, O. Y. Shuxin, and J. H. Ye, "Theoretical study of high photocatalytic performance of Ag$_3$PO$_4$," Physical Review B, vol. 83, no. 3, Article ID 035202, 2011.

33. R. Y. Liu, P. G. Hu, and S. W. Chen, "Photocatalytic activity of Ag$_3$PO$_4$ nanoparticle/TiO$_2$ nanobelt heterostructures," Applied Surface Science, vol. 258, no. 24, pp. 9805–9809, 2012. ·

34. W. G. Wang, B. Cheng, J. G. Yu, G. Liu, and W. H. Fan, "Visible-light photocatalytic activity and deactivation mechanism of Ag$_3$PO$_4$ spherical particles," Chemistry, vol. 7, no. 8, pp. 1902–1908, 2012. ·

35. Y. P. Bi, H. Y. Hu, S. X. Ouyang, G. X. Lu, J. Y. Cao, and J. H. Ye, "Photocatalytic and photoelectric properties of cubic Ag$_3$PO$_4$ sub-microcrystals with sharp corners and edges," Chemical Communications, vol. 48, no. 31, pp. 3748–3750, 2012. ·

36. M. Ge, N. Zhu, Y. P. Zhao, J. Li, and L. Liu, "Sunlight-assisted degradation of dye pollutants in Ag_3PO_4 suspension," Industrial & Engineering Chemistry Research, vol. 51, no. 14, pp. 5167–5173, 2012. ·

37. Z. B. Jiao, Y. Zhang, H. C. Yu, G. X. Lu, J. H. Ye, and Y. P. Bi, "Concave trisoctahedral Ag_3PO_4 microcrystals with high-index facets and enhanced photocatalytic properties," Chemical Communications, vol. 49, no. 6, pp. 636–638, 2013. ·

38. J. Wang, F. Teng, M. D. Chen, J. J. Xu, Y. Q. Song, and X. L. Zhou, "Facile synthesis of novel Ag_3PO_4 tetrapods and the {110} facets-dominated photocatalytic activity," Crystengcomm, vol. 15, no. 1, pp. 39–42, 2013. ·

39. C. T. Dinh, T. D. Nguyen, F. Kleitz, and T. O. Do, "Large-scale synthesis of uniform silver orthophosphate colloidal nanocrystals exhibiting high visible light photocatalytic activity," Chemical Communications, vol. 47, no. 27, pp. 7797–7799, 2011. · ·

40. J. J. Liu, X. L. Fu, S. F. Chen, and Y. F. Zhu, "Electronic structure and optical properties of Ag_3PO_4 photocatalyst calculated by hybrid density functional method," Applied Physics Letters, vol. 99, Article ID 191903, 3 pages, 2011. ·

41. Z. J. Ma, Z. G. Yi, J. Sun, and K. C. Wu, "Electronic and photocatalytic properties of Ag_3PC_4VI (C = O, S, Se): a systemic hybrid DFT study," The Journal of Physical Chemistry C, vol. 116, no. 47, pp. 25074–25080, 2012. ·

42. W. F. Yao, B. Zhang, C. P. Huang, C. Ma, X. L. Song, and Q. J. Xu, "Synthesis and characterization of high efficiency and stable Ag_3PO_4/TiO_2 visible light photocatalyst for the degradation of methylene blue and rhodamine B solutions," Journal of Materials Chemistry, vol. 22, no. 9, pp. 4050–4055, 2012. ·

43. S. B. Rawal, S. D. Sung, and W. I. Lee, "Novel Ag_3PO_4/TiO_2 composites for efficient decomposition of gaseous 2-propanol under visible-light irradiation," Catalysis Communications, vol. 17, pp. 131–135, 2012. ·

44. Y. P. Bi, S. X. Ouyang, J. Y. Cao, and J. H. Ye, "Facile synthesis of rhombic dodecahedral $AgX/Ag_3PO_4(X = Cl, Br, I)$ heterocrystals with enhanced photocatalytic properties and stabilities," Physical Chemistry Chemical Physics, vol. 13, no. 21, pp. 10071–10075, 2011. · ·

45. J. Cao, B. D. Luo, H. L. Lin, B. Y. Xu, and S. F. Chen, "Visible light photocatalytic activity enhancement and mechanism of AgBr/ Ag_3PO_4 hybrids for degradation of methyl orange," Journal of Hazardous Materials, vol. 217, pp. 107–115, 2012. ·

Optical, Electrical and Photocatalytic Properties of the Ternary Semiconductors $Zn_xCd_{1-x}S$, $Cu_xCd_{1-x}S$ and $Cu_xZn_{1-x}S$

Sandra Andrea Mayén-Hernández[1], David Santos-Cruz[1],
Francisco de Moure-Flores[1], Sergio Alfonso Pérez-García[2],
Liliana Licea-Jiménez[2], Ma. Concepción Arenas-Arrocena[3],
José de Jesús Coronel-Hernández[1], and José Santos-Cruz[1]

[1]Facultad de Química, Materiales Universidad Autónoma de Querétaro, 76010 Querétaro, QRO, Mexico

[2]Centro de Investigación en Materiales Avanzados S.C., Alianza Norte 202, 66600 Apodaca, NL, Mexico

[3]Escuela Nacional de Estudios Superiores, Unidad León, UNAM, León 36969, GTO, Mexico

ABSTRACT

The effects of vacuum annealing at different temperatures on the optical, electrical and photocatalytic properties of polycrystalline and amorphous thin films of the ternary semiconductor alloys $Zn_xCd_{1-x}S$, $Cu_xCd_{1-x}S$ and $Cu_xZn_{1-x}S$ were investigated in stacks of binary semiconductors obtained by chemical bath deposition. The electrical properties were measured at room temperature using a four-contact probe in the Van der Pauw configuration. The energy band gap of the films varied from 2.30 to 2.85 eV. The photocatalytic activity of the semiconductor thin films was evaluated by the degradation of an aqueous methylene blue solution. The thin film of $Zn_xCd_{1-x}S$ annealed under vacuum at 300°C exhibited the highest photocatalytic activity.

INTRODUCTION

Current materials science requires the creation of new, simple, and low-cost ternary and quaternary semiconductor materials with controllable chemical and physical properties. The chalcogenides have received attention for their ability to relatively easily form binary, ternary, and quaternary compounds. The physical and chemical properties of these compounds principally depend on their compositions.

The chalcogenides have many applications, including photodetectors, photovoltaic devices, optical coatings, electro-optic modulators, field-effect transistors, sensors, transducers, light sources and lasers, and photophysical and photocatalytic applications [1–16]. Ternary and quaternary compounds can easily be obtained from solid-state reactions of bilayers of binary chalcogenides because of the unique physical and chemical properties of these species, including the ionic radii of Cd^{2+}, Zn^{2+}, Cu^{2+}, and Cu^{1+}, which are 103, 83, 72, and 96 pm, respectively [17]. Additionally, wide solubility ranges can be expected for zinc and copper ions in CdS and CuS in films of ternary $Zn_xCd_{1-x}S$, $Cu_xCd_{1-x}S$ and $Cu_xZn_{1-x}S$ compounds. Zn and Cu ions can fill substitutional or interstitial sites in the lattices of CdS and CuS [18].

EXPERIMENTAL METHODS

CdS Thin Films

CdS thin films were grown on Corning glass slides by chemical bath deposition (CBD) at a temperature of $90 \pm 1°C$, over a deposition time of 40 min. The samples were prepared by immersing the substrates vertically in an aqueous solution. The reagents used to prepare the films were cadmium acetate $[Cd(CH_3COO)_2 2H_2O]$, ammonium acetate (CH_3COONH_4), ammonium hydroxide (NH_4OH), and thiourea $[(NH_2)_2CS]$. Cadmium acetate and thiourea served as the sources of Cd and S, respectively. The other components formed complexes in the reaction process and maintained a pH of 9. Highly pure water (≈ 18 MΩ) was used in the preparation of all of the solutions. The temperature was maintained at $90 \pm 1°C$ with a hot plate equipped with a magnetic stirrer. After being grown, the thin films were rinsed in highly pure water under ultrasonication for 10 min. The average thickness of the films was 80 ± 1 nm [19].

CuS Thin Films

CuS thin films were grown on CdS/Corning glass substrates by the CBD technique at $40 \pm 1°C$, over a deposition time of 60 min. The glass slides were immersed vertically in an aqueous solution of copper sulfate $(CuSO_4 \cdot 5H_2O)$, sodium acetate $(NaCOOH)$, triethanolamine $(HOCH_2CH_2)_3N$, and thiourea (CH_4N_2S). The [Cu]/[S] ratio was 1.5. Highly pure water (≈ 18 MΩ) was used in the solutions. The temperature was controlled with a hot plate equipped with a magnetic stirrer. Copper sulfate and thiourea served as the sources of copper and sulfur, respectively. After being grown, the thin films were rinsed in highly pure water under ultrasonication for 10 min [20]. The average thickness of the films was 90 ± 10 nm.

ZnS Thin Films

ZnS thin films were grown on CdS/Corning glass slides by the CBD technique at a temperature of $70 \pm 1°C$, over a deposition time of 120 min. The samples were prepared by immersing the substrates vertically in the aqueous solution. The reagents used to prepare the films were zinc acetate and thioacetamide. Zinc acetate [$(CH_3CO_2)_2Zn$] and thioacetamide (CH_3CSNH_2) served as the sources of Zn and S, respectively. Highly pure water ($\approx 18\,M\Omega$) was used in the preparation of all of the solutions. The temperature was maintained at $70 \pm 1°C$ with a hot plate equipped with a magnetic stirrer. After being grown, the thin films were rinsed in highly pure water under ultrasonication for 10 min. The average thickness of the films was 70 ± 10 nm.

Stacked bilayers of ZnS/CdS/glass, CuS/ZnS/glass, and CuS/CdS/glass were fabricated to obtain ternary compounds. After the growth of the second layer with different thicknesses, they were immediately thermally annealed at different temperatures ranging from 100 to 450°C under vacuum.

The ultraviolet-visible (UV-Vis) spectra of the films were measured on a Perkin-Elmer Lambda-2 spectrophotometer with an uncoated glass substrate placed in the reference beam. The atomic concentrations of the various elements composing the films were measured by electron dispersion spectroscopy (EDS) using a Philips XL30-ESEM. XRD measurements were performed with a RIGAKU Ultima IV using Cu-Kα radiation ($\lambda = 1.54$ Å). The films thicknesses were measured on a Sloan Dektak IIA. The resistivity of the films was characterized at ambient temperature using a Loresta-GP instrument.

A photocatalytic activity test was performed with 3.5 mL of an aqueous solution of methylene blue (MB) at a concentration of 2×10^{-5} mol/L and the ternary thin films were placed in a quartz cell with dimensions of 1 cm × 1 cm × 4 cm and irradiated with a commercial germicidal lamp ($\lambda = 252$ nm, 11 W). A rectangular sample with an area of 2 cm^2 was inserted into the interior of the quartz cell. The reaction vessel was fixed at a distance of 4.5 cm from the irradiation lamp. The irradiation times were 1 to 5 h, applied in 1-h step. The residual concentration was quantified by UV-Vis absorption spectroscopy at 663 nm. The spectrophotometer was previously calibrated with external calibration standards (2, 1.5, 1.0, 0.5, and 0.25×10^{-5} mol/L).

RESULTS AND DISCUSSION

Figure 1 shows SEM images of the ternary semiconductor compounds $Zn_xCd_{1-x}S$, $Cu_xCd_{1-x}S$ and $Cu_xZn_{1-x}S$ annealed under vacuum at 250°C and 450°C. All films covered the substrate surface and were free of pinholes. For films of the semiconductor compound $Zn_xCd_{1-x}S$ annealed at 250°C, the superficial morphology was composed of agglomerates approximately 160 nm in diameter with a well-defined spherical shape (Figure1(a)). The spheres contained small particles measuring 8.5 nm. The spherical shapes of the agglomerates were lost in films annealed at 450°C and nanocracks are evident on the surface of these films. The $Cu_xCd_{1-x}S$ thin films annealed at 250°C were composed of irregular agglomerates with approximately circular, asymmetric shapes, and a distribution of sizes and both small particles (Figure 1(c)) and nanocracks were evident. At an annealing temperature of 450°C, the agglomerates exhibited better-defined shapes and stacking faults (Figure1(d)).The thin films of the $Zn_xCd_{1-x}S$ semiconductor demonstrated a superficially amorphous nature (Figures1(e) and 1(f)).

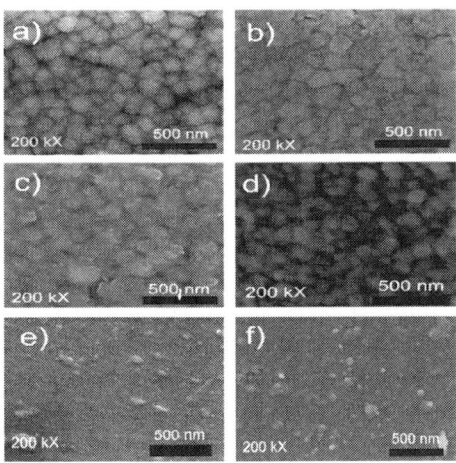

Figure 1: SEM images of the chalcogenides studied (a) and (b) $Zn_xCd_{1-x}S$, (c) and (d) $Cu_xCd_{1-x}S$, and (e) and (f) $Cu_xZn_{1-x}S$ annealed at 250°C ((a), (c), and (e)) and 450° ((b), (d), and (f)), respectively.

Table 1 shows the EDS analysis results of the three ternary compounds at the vacuum annealing temperatures of 250 and 450°C.

The compounds containing cadmium are generally rich in cadmium at low temperature (250°C); however, at high temperature, where diffusion is favored, the cadmium concentration diminishes and, consequently, the respective concentrations of zinc and copper increase ($Zn_xCd_{1-x}S$ and $Cu_xCd_{1-x}S$). The compound $Cu_xZn_{1-x}S$ is rich in zinc at low temperature. The copper concentration increases at high temperature.

Table 1: Mean compositions in at. %, determined by EDS of the (vacuum-annealed) ternary semiconductor compounds

Sample	$T_{Anm}/(°C)$	Zn (at.%)	Compound
$Zn_xCd_{1-x}S$	250	29.43	$Zn_{0.3}Cd_{0.7}S$
	450	44.84	$Zn_{0.45}Cd_{0.55}S$
		Cu (at. %)	
$Cu_xCd_{1-x}S$	250	11.24	$Cu_{0.11}Cd_{0.89}S$
	450	18.03	$Cu_{0.18}Cd_{0.82}S$
$Cu_xZn_{1-x}S$		Cu (at. %)	
	250	28.59	$Cu_{0.29}Zn_{0.71}S$
	450	34.6	$Cu_{0.35}Zn_{0.65}S$

Figures 2(a) and 2(b) display the XRD results obtained over a 2θ range of 15 to 60 degrees for the ternary semiconductor alloys, $Cu_xCd_{1-x}S$ $Cu_xZn_{1-x}S$ and $Zn_xCd_{1-x}S$ annealed under vacuum. For Figure 2(a) in all cases, the thin films were amorphous independent of the annealing temperature. At 250°C, the samples were most amorphous, and at 450°C, small peaks appeared. Figure 2(a) (A) shows the XRD results obtained for thin films of $Cu_xCd_{1-x}S$, the patterns show two peaks at 44.28° and 52.44° and at 43.04° and 47.8° for films annealed under vacuum at 250°C and 450°C, respectively. Figure 2(a) (B) shows the XRD results obtained for nanocrystalline thin films of $Cu_xZn_{1-x}S$, which indicate a single one peak at 43.26° and 45.16° for films annealed under vacuum at 250°C and 450°C, respectively. The small peaks made it difficult to identify the phase.

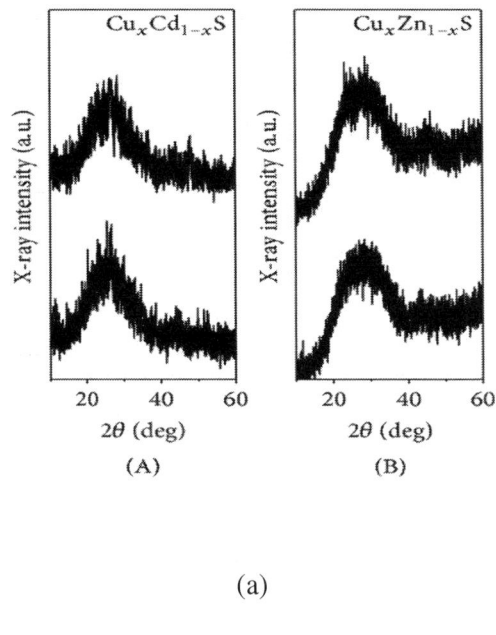

(a)

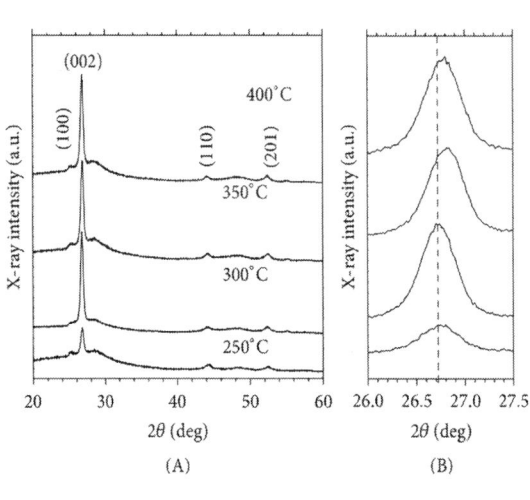

(b)

Figure 2: (a) X-ray diffraction patterns of the ternary semiconductor compounds annealed at 250°C (bottom patterns) and 450°C (top patterns) of (A) $Cu_xCd_{1-x}S$ and (B) $Cu_xZn_{1-x}S$. (b) X-ray diffraction patterns of the ternary semiconductor compounds of $Zn_xCd_{1-x}S$, (A) as a function of annealing temperature and (B) zoom in of the peaks (002) to observe the shift.

As shown in Figure 2(b) (A), at 250–450°C $Zn_xCd_{1-x}S$ shows peaks at 25.13°, 26.8°, 44.2°, and 52.4°, corresponding to the hexagonal wurtzite planes of (100), (002), (110), and (201) in thin films of the standard PDF Card 00-049-1302. Figure 2(b) (B) shows that the orientation of the films is in the plane (002), with the increase of annealing temperature, these peaks move to higher diffraction angle, indicative of an increase of Zn content in the films. The grains size was calculated from Scherrer's equation for the peaks (002) as a function of the annealing temperature and varied for 16.316 nm, 19.95 nm, 19.43 nm, and 19.50 nm, respectively, for 250, 300, 350, and 400°C.

Figures 3(a), 3(b), and 3(c), respectively, show the UV-Vis transmission spectra as a function of annealing temperature for $Zn_xCd_{1-x}S$, $Cu_xCd_{1-x}S$ and $Cu_xZn_{1-x}S$ films annealed under vacuum. The $Zn_xCd_{1-x}S$ films exhibited high transmittances > 80% over an annealing temperature range of 250°C to 450°C. For the $Cu_xCd_{1-x}S$ thin films, the transmittance was approximately 65%; higher transmittance was observed for the films annealed under vacuum at 300°C than for those annealed at 250°C. The $Cu_xZn_{1-x}S$ thin films showed the lowest transmittance values of 35 to 50%; the highest transmittance was observed for the films annealed at 350°C and the lowest for those annealed at 300°C. The optical band gap, E_g, of each film was calculated using the Tauc equation by plotting $(\alpha h v)^2$ as a function of hv, where α is the optical absorption coefficient and hv is the photon energy. In the graphs, the linear portion of the curve was extrapolated to $(\alpha h v)^2. = 0$ The band gap is shown as a function of the annealing temperature in Figure 4. For $Zn_xCd_{1-x}S$, the E_g values decreased from 2.83 to 2.5 eV as a function of the vacuum annealing temperature. For the $Cu_xCd_{1-x}S$ thin films, the E_g values decreased from 2.43 to 2.28 eV by increasing the annealing temperature. The $Zn_xCd_{1-x}S$ thin films showed values of 2.80 to 2.65 eV, with the increase of the annealing temperature. The E_g values are in agreement with the values reported by other research groups [21–23]. In general, the band gap decreases as the annealing temperature increases; this effect may be caused by both the greater diffusion between the bilayers and the formation of the ternary alloy.

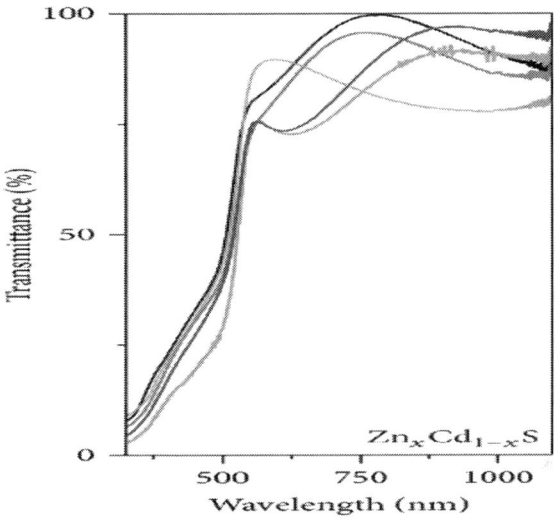

(a)

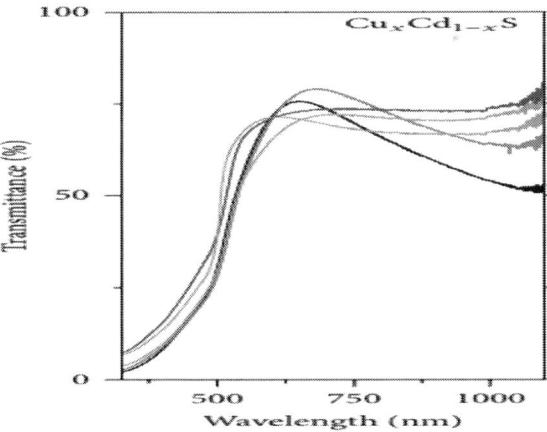

(b)

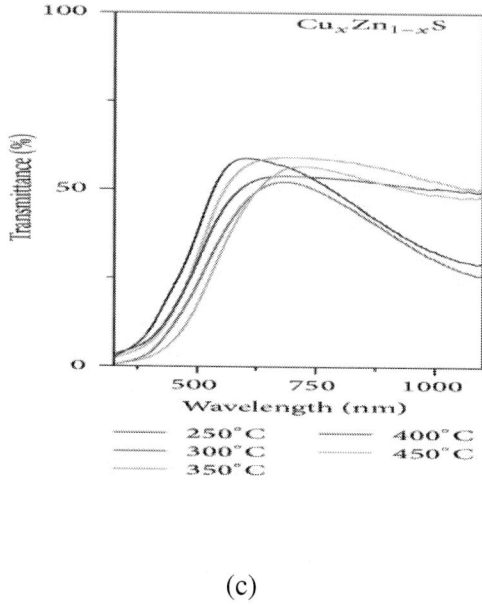

(c)

Figure 3: Optical transmittance spectra of the ternary semiconductor compounds studied.

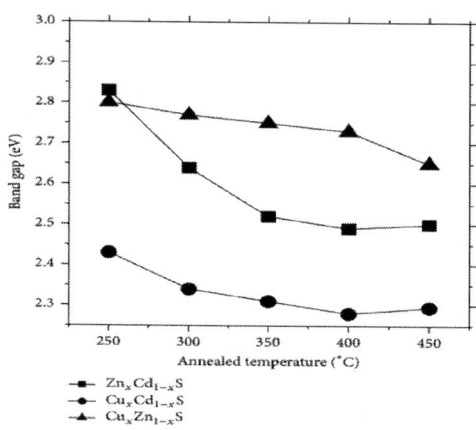

Figure 4: Band gap results as a function of the annealing temperature.

Figure 5 shows the resistivity values of the ternary semiconductor compounds. The thin films of $Zn_xCd_{1-x}S$ were highly resistive independent of thermal annealing; the resistivity values were greater than $10^6\,\Omega\cdot$cm (results not shown). Figure 5 displays the resistivity values of the $Cu_xCd_{1-x}S$ and $Cu_xZn_{1-x}S$ films annealed under vacuum as a function of the annealing temperature. The values for $Cu_xCd_{1-x}S$ decreased as the annealing temperature increased. The minimum resistivity occurred at 300°C ($2.74 \times 10^{-3}\,\Omega\cdot$cm), above this temperature, the resistivity value increased. The resistivity values for the $Cu_xZn_{1-x}S$ samples decreased as the annealing temperature increased. The lowest value was obtained at 350°C ($4.11 \times 10^{-4}\,\Omega\cdot$cm) and, for temperatures above 350°C, the resistivity values increased.

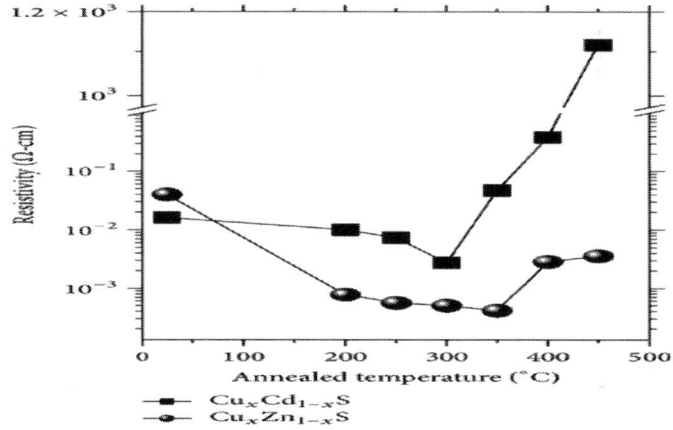

Figure 5: Resistivity values as a function of the annealing temperature.

The change of resistivity values with the change of the annealing temperatures could be attributed to the diffusion of copper ions on the matrix of CdS and ZnS, with the temperature increase and the presence of nanocracks on the surface of the films at elevated temperatures. The ionic radii of Cd^{2+}, Zn^{2+}, and Cu^{2+}, are 103, 83, and 72 pm, respectively [17], on the other hand, the copper ions could be introduced more effectively with the increase of the temperature until the saturation, and possibly the segregation phenomena takes place at different temperatures (300 and 350°C for $Cu_xCd_{1-x}S$ and $Cu_xZn_{1-x}S$, resp.). Copper ions are closer to the ionic radii of Zn^{2+} than Cd^{2+}. Additionally, copper ions could be introduced more easily in the ZnS lattice occupied substitutional or interstitial sites in the lattices of ZnS than in CdS [18].

To demonstrate the prospective photocatalytic applications of the fabricated materials, the thin films were used as catalysts in the degradation of methylene blue (MB) under germicidal irradiation (λ = 252 nm) at room temperature. Figure 6 displays the normalized concentration of MB in aqueous solution as a function of the vacuum annealing temperature for the three different semiconductor alloys. The concentration was quantified based on the absorption at 653 nm.

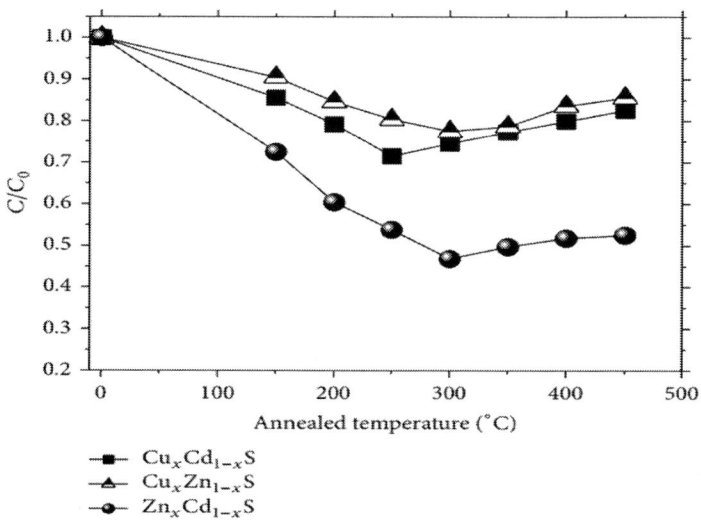

Figure 6: Photobleaching of the normalized concentration of MB after 3 h as a function of the annealing temperature.

The absorption is time-dependent, initially, the irradiation time was held constant at 3 h for all samples. The ternary semiconductor samples generally showed photocatalytic behavior. The sample with the best photobleaching performance was $Zn_xCd_{1-x}S$ annealed at 300°C (see Figure 6).

As indicated in Figure 6, the best samples were $Zn_xCd_{1-x}S$ and $Cu_xCd_{1-x}S$ annealed at 300°C and 250°C, respectively. The temporal dependence of the photobleaching of MB by these samples was analyzed. Figure7(a) shows the normalized MB concentration as a function of time in hours. The $Zn_xCd_{1-x}S$ film demonstrated that it is an excellent candidate for photocatalytic applications. In 5 h, the mineralization of MB was nearly complete (82%) for the thin films of

$Zn_xCd_{1-x}S$ annealed under vacuum at 300°C. This can be attributed to the synergistic effect of thermal annealing, better crystallinity, and concentration of the Zn mole fraction to 0.3 and 0.33 for 250 and 300°C, respectively; similar results were found by an optimal Zn incorporation corresponding to x approximately 0.2–0.33 [24–27]. Additionally these results are comparable to the photocatalytic activities reported in the literature for thin films of other materials, such as ZnO and TiO_2, which have been shown to be excellent catalyzers [28–31].

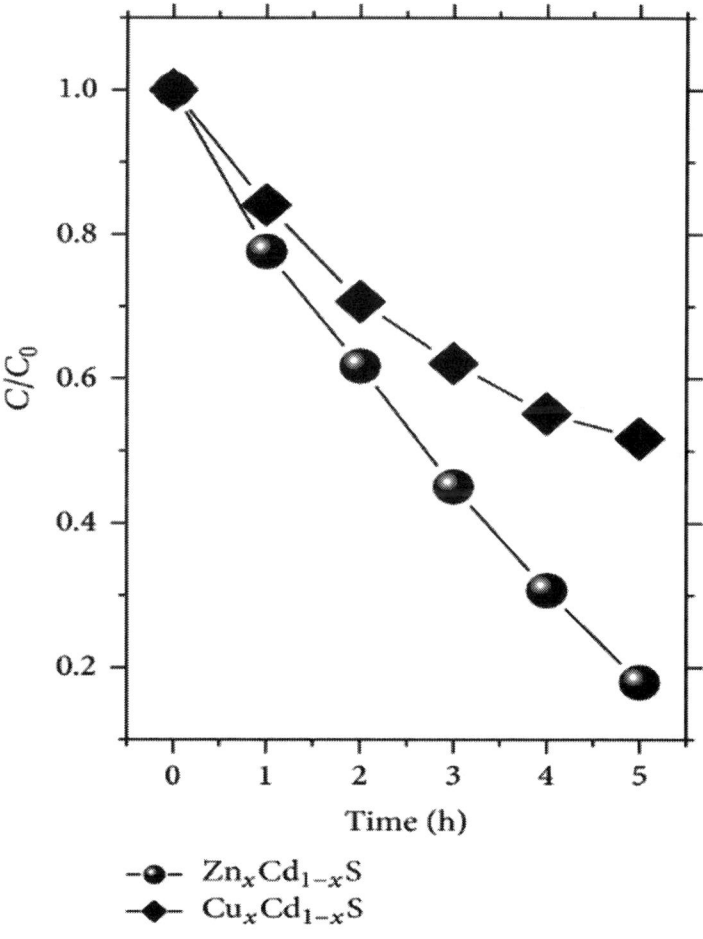

(a)

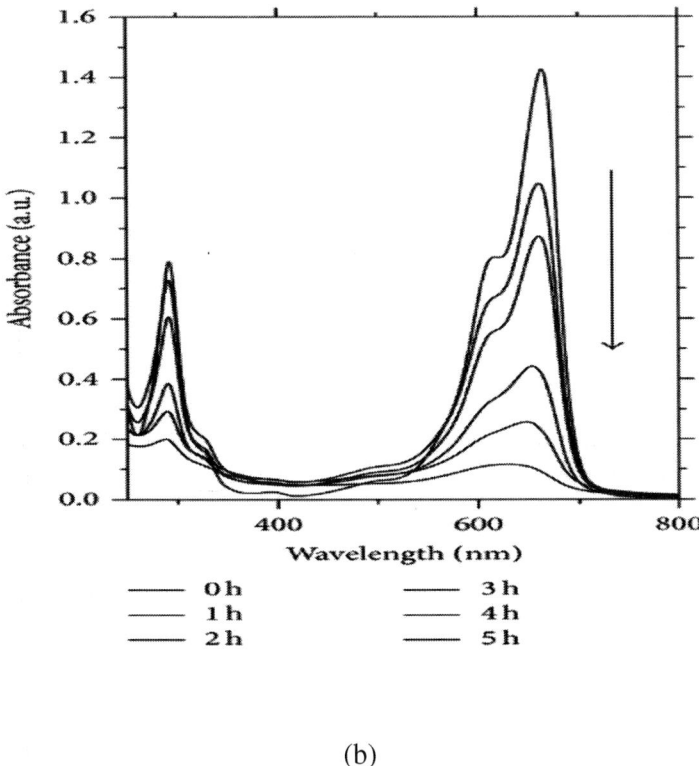

(b)

Figure 7: (a) Photobleaching of the normalized concentration of MB as a function of time for $Zn_xCd_{1-x}S$ and $Cu_xCd_{1-x}S$. (b) Absorption spectra of the $Zn_xCd_{1-x}S$ thin films.

The thin films of $Zn_xCd_{1-x}S$ were obtained by chemical bath deposition and were annealed under vacuum at 300°C. Obtaining this film is very economical compared to the preparation of films of ZnO (sol-gel, sputtering at T≥500°C) and TiO_2 (sol-gel, sputtering at T≥450°C). Additionally, our films have high transmittances (≥85%) and band gaps (2.5 to 2.83 eV) closer to the absorption spectrum of visible light than gaps larger than 3.2 eV (ZnO and TiO_2).

For $Cu_xCd_{1-x}S$ annealed at 250°C, the extent of mineralization after 5 h was approximately 50%. Figure 7(b) displays the absorption spectra of an aqueous solution of MB at different intervals in the presence of a sample of $Zn_xCd_{1-x}S$ annealed at 300°C. The concentration was quantified based on the absorption at 653 nm. The absorption was time-dependent and the complete mineralization of MB was nearly reached after 5 h under the experimental conditions described above.

CONCLUSIONS

Simple and economical syntheses of ternary semiconductor alloys of $Zn_xCd_{1-x}S$, $Cu_xCd_{1-x}S$ and $Cu_xZn_{1-x}S$ were obtained by the chemical bath deposition and subsequent annealing of stacked bilayers of binary semiconductor films. The three compounds showed good optical properties, superficial morphology, and photocatalytic activity. The band gap of the $Cu_xCd_{1-x}S$ and $Cu_xZn_{1-x}S$ samples was tuned from 2.3 to 2.85 eV. These samples demonstrated the lowest resistivity values of 2.74×10^{-3} and 4.11×10^{-4} $\Omega \cdot$cm, respectively, among all of the films that were fabricated. Extensive photodegradation of methylene blue was observed for the $Zn_xCd_{1-x}S$ film annealed at 300°C in vacuum; this activity is comparable to that reported in the literature for thin films of ZnO and TiO_2.

ACKNOWLEDGMENTS

The authors thank the Universidad Autónoma de Querétaro México for the financial support of this work through the Project FOFI 2012 and the CONACYT Projects CB154787 and CB176450.

REFERENCES

1.   K. T. R. Reddy and P. J. Reddy, "Studies of $Zn_xCd_{1-x}S$ films and $Zn_xCd_{1-x}S$/CuGaSe$_2$ heterojunction solar cells," Journal of Physics D: Applied Physics, vol. 25, no. 9, pp. 1345–1348, 1992.

2.   R. N. Bhattacharya, M. A. Contreras, B. Egaas, R. N. Noufi, A. Kanevce, and J. R. Sites, "High efficiency thin-film $CuIn_{1-x}Ga_xSe_2$ photovoltaic cells using a $Cd_{1-x}Zn_xS$ buffer layer," Applied Physics Letters, vol. 89, no. 25, Article ID 253503, 2006.

3.   I. O. Oladeji, L. Chow, C. S. Ferekides, V. Viswanathan, and Z. Zhao, "Metal/CdTe/CdS/$Zn_xCd_{1-x}S$/TCO/glass: a new CdTe thin

film solar cell structure," Solar Energy Materials Solar Cells, vol. 61, no. 2, pp. 203–211, 2000.

4. I. O. Oladeji and L. Chow, "Synthesis and processing of CdS/ZnS multilayer films for solar cell application," Thin Solid Films, vol. 474, no. 1-2, pp. 77–83, 2005.

5. R. Nagarajan, "Nanoparticles: building blocks for nanotechnology," in Nanoparticles: Synthesis, Stabilization, Passivation, and Functionalization, vol. 996 of ACS Symposium Series, chapter 1, pp. 2–14, American Chemical Society, 2008.

6. T. L. Chu, S. S. Chu, J. Britt, C. Ferekides, and C. Q. Wu, "Cadmium zinc sulfide films and heterojunctions," Journal of Applied Physics, vol. 70, no. 5, pp. 2688–2693, 1991.

7. B. Kumar, P. Vasekar, S. A. Pethe, N. G. Dhere, and G. T. Koishiyev, "Zn_xCd_{1-x} as a heterojunction partner for $CuIn_{1-x}Ga_xS_2$ thin film solar cells," Thin Solid Films, vol. 517, no. 7, pp. 2295–2299, 2009.

8. C. Jin, W. Zhong, X. Zhang, Y. Deng, C. Au, and Y. Du, "Synthesis and wavelength-tunable luminescence property of wurtzite Zn xCd1-xS nstructures," Crystal Growth and Design, vol. 9, no. 11, pp. 4602–4606, 2009.

9. F. Benkabou, H. Aourag, and M. Certier, "Atomistic study of zinc-blende CdS, CdSe, ZnS, and ZnSe from molecular dynamics," Materials Chemistry and Physics, vol. 66, no. 1, pp. 10–16, 2000.

10. H. E. Ruda, Wide-Gap II–VI Compounds for Opto-Electronic Applications, Springer, 1992.

11. T. P. Kumar, S. Saravanakumar, and K. Sankaranarayanan, "Effect of annealing on the surface and band gap alignment of CdZnS thin films," Applied Surface Science, vol. 257, no. 6, pp. 1923–1927, 2011.

12. N. A. Noor, N. Ikram, S. Ali, S. Nazir, S. M. Alay-E-Abbas, and A. Shaukat, "First-principles calculations of structural, electronic and optical properties of $Cd_xZn_{1-x}S$ alloys," Journal of Alloys and Compounds, vol. 507, no. 2, pp. 356–363, 2010.

13. M. Nyman, K. Jenkins, M. J. Hampden-Smith et al., "Feed-Rate-Limited Aerosol-Assisted Chemical Vapor deposition of $Cd_xZn_{1-x}S$ and ZnS:Mn with Composition Control," Chemistry of Materials, vol. 10, no. 3, pp. 914–921, 1998.

14. P. Chen, J. E. Nicholls, M. O'Neill et al., "Spectroscopic evidence for different laser gain mechanisms in optically pumped ZnCdS/ZnS quantum well structures," Journal of Applied Physics, vol. 84, no. 10, pp. 5621–5625, 1998.

15. S. A. Al Kuhaimi and Z. Tulbah, "Structural, compositional, optical, and electrical properties of solution-grown ZnxCd1-xS films," Journal of the Electrochemical Society, vol. 147, no. 1, pp. 214–218, 2000.

16. H. Liu and Y. Zhu, "Synthesis and characterization of ternary chalcogenide ZnCdS 1D nanostructures,"Materials Letters, vol. 62, no. 2, pp. 255–257, 2008.

17. R. Chang, Chemistry, McGraw Hill, 9th edition, 2007.

18. A. A. Ziabari and F. E. Ghodsi, "Influence of Cu doping and post-heat treatment on the microstructure, optical properties and photoluminescence features of sol-gel derived nanostructured CdS thin films,"Journal of Luminescence, vol. 141, pp. 121–129, 2013.

19. J. S. Cruz, R. C. Pérez, G. T. Delgado, and O. Z. Angel, "CdS thin films doped with metal-organic salts using chemical bath deposition," Thin Solid Films, vol. 518, no. 7, pp. 1791–1795, 2010.

20. J. Santos Cruz, S. A. Mayén Hernández, R. Mejía Rodríguez, R. Castanedo Pérez, G. Torres Delgado, and S. Jiménez Sandoval, "Effect of the sintering temperature on the photocatalytic activity of CdO+ CdTiO 3 thin films," Chalcogenide Letters, vol. 9, no. 2, pp. 85–91, 2012.

21. M. A. Mahdi, J. J. Hassan, Z. Hassan, and S. S. Ng, "Growth and characterization of $Zn_xCd_{1-x}S$ nanoflowers by microwave-assisted chemical bath deposition," Journal of Alloys and Compounds, vol. 541, pp. 227–233, 2012.

22. M. Adelifard, H. Eshghi, and M. M. B. Mohagheghi, "Synthesis and characterization of nanostructural CuS-ZnS binary compound thin films prepared by spray pyrolysis," Optics Communications, vol. 285, no. 21-22, pp. 4400–4404, 2012.

23. A. Abdolahzadeh Ziabari and F. E. Ghodsi, "Effects of the Cd:Zn:S molar ratio and heat treatment on the optical and photoluminescence properties of nanocrystalline CdZnS thin

films," Materials Science in Semiconductor Processing, vol. 16, no. 6, pp. 1629–1636, 2013.

24. X. Xu, R. Lu, X. Zhao, Y. Zhu, S. Xu, and F. Zhang, "Novel mesoporous $Zn_xCd_{1-x}S$ nanoparticles as highly efficient photocatalysts," Applied Catalysis B: Environmental, vol. 125, pp. 11–20, 2012.

25. J. Ran, J. Yu, and M. Jaroniec, "$Ni(OH)_2$ modified CdS nanorods for highly efficient visible-light-driven photocatalytic H_2 generation," Green Chemistry, vol. 13, no. 10, pp. 2708–2713, 2011.

26. J. Zhang, L. F. Qi, J. R. Ran, J. G. Yu, and S. Z. Qiao, "Ternary NiS/ $Zn_xCd_{1-x}S$/reduced graphene oxide nanocomposite for enhanced solar photocatalytic H_2-production activity," Advanced Energy Materials, 2014.

27. J. R. Ran, J. Zhang, J. G. Yu, M. Jaroniec, and S. Z. Qiao, "Earth-abundant cocatalysts for semiconductor-based photocatalytic water splitting," Chemical Society Reviews, 2014.

28. G. Torres Delgado, C. I. Zúñiga Romero, S. A. Mayén Hernández, R. Castanedo Pérez, and O. Zelaya Angel, "Optical and structural properties of the sol-gel-prepared ZnO thin films and their effect on the photocatalytic activity," Solar Energy Materials & Solar Cells, vol. 93, no. 1, pp. 55–59, 2009.

29. A. Enesca, L. Isac, and A. Duta, "Hybrid structure comprised of SnO_2, ZnO and Cu_2S thin film semiconductors with controlled optoelectric and photocatalytic properties," Thin Solid Films, vol. 542, pp. 31–37, 2013.

30. R. A. Carcel, L. Andronic, and A. Duta, "Photocatalytic activity and stability of TiO_2 and WO_3 thin films," Materials Characterization, vol. 70, pp. 68–73, 2012.

31. H. Yu and C. Wang, "Photocatalysis and characterization of the gel-derived TiO_2 and $P-TiO_2$transparent thin films," Thin Solid Films, vol. 519, no. 19, pp. 6453–6458, 2011.

Chapter 3

Recent Advances in Heterogeneous Photocatalytic Decolorization of Synthetic Dyes

Nurhidayatullaili Muhd Julkapli, Samira Bagheri,
and Sharifah Bee Abd Hamid

Nanotechnology & Catalysis Research Centre (NANOCAT), IPS
Building, University Malaya, 50603 Kuala Lumpur, Malaysia

ABSTRACT

During the process and operation of the dyes, the wastes produced were commonly found to contain organic and inorganic impurities leading to risks in the ecosystem and biodiversity with the resultant impact on the environment. Improper effluent disposal in aqueous ecosystems leads to reduction of sunlight penetration which in turn diminishes photosynthetic activity, resulting in acute toxic effects on the aquatic flora/fauna and dissolved oxygen concentration. Recently, photodegradation of various synthetic dyes has been studied in terms of their absorbance and the reduction of oxygen content by changes in the concentration of the dye. The advantages that make photocatalytic techniques superior to traditional methods are the ability to remove contaminates in the range of ppb, no generation of polycyclic compounds, higher speed, and lower cost. Semiconductor metal oxides, typically TiO_2, ZnO, SnO, NiO, Cu_2O, Fe_3O_4, and also CdS have been utilized as photocatalyst for their nontoxic nature, high photosensitivity, wide band gap and high stability. Various process parameters like photocatalyst dose, pH and initial dye concentrations have been varied and highlighted. Research focused on surface modification of semiconductors and mixed oxide semiconductors by doping them with noble metals (Pt, Pd, Au, and Ag) and organic matter (C, N, Cl, and F) showed enhanced dye degradation compared to corresponding native semiconductors. This paper reviews recent advances in heterogeneous photocatalytic decolorization for the removal of synthetic dyes from water and wastewater. Thus, the main core highlighted in this paper is the critical selection of semiconductors for photocatalysis based on the chemical, physical, and selective nature of the poisoning dyes.

INTRODUCTION

Photocatalytic Decolorization in Water and Wastewater Treatment

Generally, dyes are complex unsaturated aromatic compounds with accomplishing characteristics like color, intensity, solubility, fastness, and substantiveness [1, 2]. It could be compounds with different coloring particles, each varying in type from each other in terms of chemical composition, and are used for coloring textiles in different colors and shades that are completely soluble in aqueous media [2, 3]. Dyes derived from inorganic or organic compounds are called synthetic dyes and they are categorized based on their basic chemistry (Table 1; Figure 1). There are various ways used for the assortment of dyes. It should be noted that each category of dyes has an exclusive chemistry, source of materials, nature of its respective chromophores, nuclear structure, industrial classification, and specific way of bonding.

Although some dyes can chemically react with the substrates forming robust bonds in the process, others can be sustained by physical forces. The most common synthetic dyes in use today are dispersible types for polyester dyeing and reactive and direct types for cotton dyeing.

Table 1: Usage and characterization of dyes

Group of dyes	Characteristics	Application	References
Direct dyes	(i) Dyeing process with one action, without the assistance of an affixing agent; simplest and cheapest dyes (ii) Water soluble anionic dyes; substantive to form aqueous media in the electrolytes (iii) High affinity for cellulose fibers (iv) Apply to the dye materials to improve wash fastness properties (chelation with salts of metals and treatment with a cationic dye-complexing resin or formaldehyde) (v) Some contain sulphonate functionality to improve solubility (negative charge of dyes and fibers repel each other) (vi) Its flat length enable and shape to lie along-side cellulose fiber and maximum (vii) Van-der-Waals, hydrogen bonds, and dipole (ix) Dyeing method: exhaust/beck/continuous	Cotton, cellulosic, regenerated cellulose, paper, leather, nylon, and blends	[6, 7]
Vat dyes	(i) Water insoluble dyes (ii) Apply as soluble leuco salt after reduction in an alkaline solution with sodium hydrogen sulfide (iii) The leuco form is reoxidized to the insoluble keto form to redevelop the crystal structure (iv) More chemically complex (v) Dyeing methods: exhaust, package, continuous	Cotton, linen and rayon, soap	[8, 9]
Organic pigments	(i) Negatively charged compounds (ii) Made from ground up colored rocks, minerals, animals, and plants (iii) No chemical information (iv) Classification based on the dye's source and color (v) Application requires a mordant	Cotton, paper, cellulosic, blended fabrics	[10, 11]
Reactive dyes	(i) React directly with the fiber molecules to form chemical bonds (ii) Conceivable to achieve very high wash fastness properties (iii) Require facile dyeing methods (iv) Simple chemical structure (v) The largest dye class (vi) Adsorption spectra with a narrow adsorption band (vii) Dyeing is bright (viii) Dyeing methods: exhaust, beck cold pad batch, and continuous	Cellulosic fabric and fibers	[12, 13]
Dispersed dyes	(i) Water insoluble nonionic (ii) Require additional factors (dye carrier, pressure, and heat) to penetrate synthetic dyes (iii) Dispersed in aqueous media wherever the dye is dissolved into fibers (iv) Especially on polyester and to a lesser extent on cellulose acetate, nylon, acrylic fibers, and cellulose (v) Niche market in dye diffusion thermal transfer process for electronic photography and thermal transfer printing (vi) Dyeing method: high temperature exhaust, continuous	Synthetic/ hydrophobic fibers from aqueous dispersion	[14, 15]

Acid dyes	(i) Water soluble anionic dyes (ii) Typical pollutants: color, organic acid, unfixed dyes (iii) Dyeing methods: exhaust, beck, and continuous	Silk, wool, synthetic fibers, leather, nylon, modified acrylics, paper, ink-jet printing, food, cosmetics	[16, 17]
Azoic dyes	(i) Contain one azo group (mono azo), two azo group (disazo), three azo group (trisazo), four azo group (tetrakisazo), or more (polyazo) groups (ii) Attach to two classes of which at least one but usually both are aromatic (iii) Exist in the transform 1 in (iv) which the bond angle is 120° and the nitrogen atoms are sp^2 hybridized (v) Consist of electron accepting substituents and electron donating substituents (vi) Named as carbocyclic azo dyes if include only aromatic groups (naphthalene and benzene) (vii) Named as heterocyclic azo dyes if include heterocyclic group	Printing inks, pigments	[18, 19]
Basic dyes	(i) Water soluble cationic dyes (ii) Can be applied directly to cellulosic with no mordants (or metal-like copper and chromium) (iii) Yield colored cations in solutions (iv) Apply as brightness of shade is more important than fastness to washing and light (v) Some basic dyes show biological activity and are used in medicine as antiseptics (vi) Salt-forming counter ion (vii) Colorless anion of a low molecular mass, organic, or inorganic acid (viii) Can be turned to water soluble dye bases by addition of alkali (ix) The positive charge is localized on an ammonium group (x) Dyeing methods: exhaust, beck, and continuous	Silk, wool, cotton, polyacrylonitrile, modified nylons, modified polyester, tannin-mordanted cotton	[20, 21]
Oxidation dyes	(i) Primarily aromatic compounds that belong to three major chemical families (Diamines, Aminophenols (amino naphthols) and Phenols or naphthols) (ii) Colorless and are typically a low molecular weight product (iii) Categories-oxidation base as a primary, intermediate and coupler as a secondary, intermediate	Hair	[22, 23]
Developed dyes	Any group of direct azo dyes which after applying to the fiber can be diazotized further and coupled on the fiber to form shades faster to washing	Cellulosic fibers, fabric	[24, 25]
Mordant dyes	A substance utilized to set dyes on fabrics or tissue sections by forming a coordination complex with the dye that attaches to the tissue or fabric	Cellulosic fibers, fabric, silk, wool	[26–28]
Optical/ fluorescent brightener	(i) Absorb light in the violet region and ultraviolet (mostly 340–370 nm) of the electromagnetic spectrum, and reemit light in the blue region (usually 420–470 nm) (ii) Utilized to increase the appearance of color of paper and fabric, causing a "whitening" effect, making materials look less yellow by increasing the overall amount of blue light reflected	Synthetic fibers, leather, cotton, sport goods	[29–31]

Solvent dyes	(i) Water insoluble (ii) Free of polar solubilizing groups such as carboxylic acid, sulfonic acid, or quaternary ammonium	Wood staining, solvent inks, waxes, coloring oils, plastic, gasoline oil	[32, 33]
Anthraquinone	(i) The oldest dyes (4000 years) (ii) No natural counterpart (iii) Low cost effectiveness	Wrapping of mummies	[34]
Indigoid	(i) Expensive (ii) Made of tyrian purple (iii) Give progressively paler blue shades (iv) Oxidation process of indigoid gives phenylacetic acid	Textile, wool, linen, cotton Use exclusively for dyeing denim jeans, jackets	[35]
Sulfur dyes	(i) Made by heating aromatic or heterocyclic compounds with species that release sulfur or sulfur (ii) Classified by sulfur bake, polysulfide melt dyes, and polysulfide bake (iii) Not well-defined chemical compounds (iv) Mostly contain various thiophenolic and heterocyclic sulfurs (v) On oxidation, the monomeric molecules cross-linked into large molecules form disulfide bridge (vi) Dyeing methods: continuous	Cotton, other cellulosic	[36]

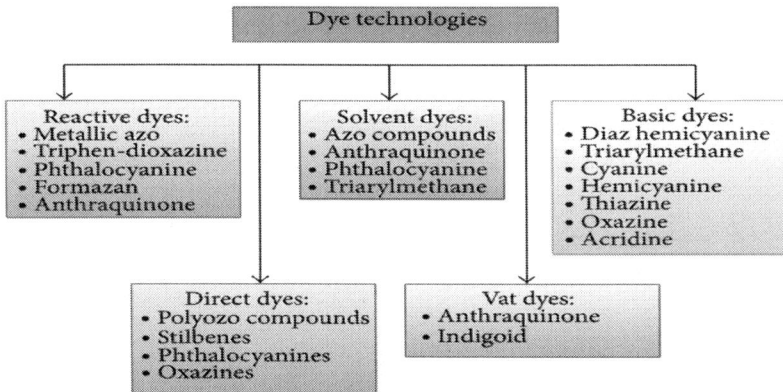

Figure 1: Synthetic dyes and its derivatives.

Synthetic dyes are also utilized in high technology applications, like in the electronics, medical, and specifically the nonimpact printing industries. For instance, they are utilized in electrophotography (laser printing and photocopying) in both the organic photoconductor and the toner, in direct and thermal transfer printing, and also in ink-jet printing. With increasing synthetic dye usage, dye removal becomes an important but challenging area of research for wastewater treatment since most of dyes and their degradation products may be carcinogenic and toxic to mammals [4, 5].

Heterogenous photocatalysis using semiconductors for water and wastewater treatment continues to attract much interest [4, 5, 37, 38]. The lower cost of catalysts and the utilization of environmental protection and renewable energy form this technology to be adequately attractive compared to other techniques [37]. Because the process relies on the photoactivation of semiconductors, the efficiency of the catalyst is qualified by the capacity to generate electron-hole pairs in addition to radical production [39, 40]. Hence, the selection of proportionate semiconductors is the key to reactivity control [38].

Poisoning Dyes

Only 45 to 47% of dyes have been reported as organic dyes with biodegradable and solubility characteristics. The remaining 55 to 53% of dyes are toxic and their persistence in wastewater has recently become an issue of interest [8, 41, 42]. Synthetic or poisoning dyes engaged more often on industrial scale are acid dyes, water soluble anionic, basic dyes-water soluble cationic, substantive dyes-alkaline, vat dyes-water soluble alkali metal salt, azoic dyes, sulfur dyes, and chrome dyes. Generally, there are two important components in the dye molecules: chromophore component that is responsible for producing the color and the auxochrome component which increases the affinity of the dye towards cellulose fibers [14, 43].

The mentioned dyes are released in aqueous streams as effluents of several industries, including textiles, paper, leather, plastic, automobile, furniture, finishing sector, and others, which consequently create intense environmental pollution problems via the release of potential carcinogenic and toxic substances into the aqueous phase [22, 23]. The discharge of an enormous volume of wastewater containing dyes is an inevitable consequence, because the textile industry consumes large quantities of water and all dyes cannot be completely combined with fibers during the dyeing process. More than 79105 metric tonnes of dye stuffs are produced worldwide annually, with 10 to 50% of this amount being released into wastewater [18, 44]. These high concentrations of dyes in effluents interfere with the penetration of visible light into the water, resulting in a hindrance to photosynthesis and a decrease in gas solubility, since less than 1 mgL^{-1} of dye is highly visible. Furthermore, synthetic dyes, which include an aromatic ring in their basic structure, are regarded as toxic, carcinogenic, and xenobiotic compounds [43–46]. Also, this type of dyes may convey toxicity to aquatic life and may be mutagenic and carcinogenic and can cause intense damage to human beings, including the reproductive system and dysfunction of the kidneys, brain, liver, and central nervous system [34].

Therefore, decolorization and detoxification of dye-containing wastewater need to be conducted before discharging wastewater into natural water bodies [26, 27, 29]. Certain physical, chemical, and biological treatments are currently being used for dye wastewater treatment. Although physical and chemical methods usually show high dye-removal efficiencies, high operating costs are the main drawback due to the large-scale application of these methods [32, 47, 48]. Furthermore, due to the high chemical stability of synthetic dyes, conventional biological treatment using bacteria cannot remove the dyes efficiently [43, 49,50].

PHOTOCATALYTIC DECOLORIZATION OF SYNTHETIC DYES

The complete degradation of the dyes is not possible by conventional methods such as precipitation, adsorption, flocculation, flotation, oxidation, reduction, electrochemical, aerobic, anaerobic, and biological treatment methods. These methods have inherent limitations in technologies such as less efficiency and production of secondary sludge, the disposal of which is a costly affair [43–53]. Merely, transferring hazardous materials from one medium to another is not a long-term solution to the issue of toxic waste loading on the environment [30]. Many technologies have been applied to remedy dyes from wastewater, like coagulation/flocculation, biological treatment, electrochemical, membrane filtration, ion exchange, adsorption, and chemical oxidation [54, 55]. Chemical coagulations for dye removal require loading of chemical coagulation and optimal operating conditions like pH and coagulation dosage should be rigidly reminded for achieving maximum dye removal [56]. The coagulation-flocculation process can be utilized as a pre- or post- or even as a main treatment. This process is cost effective and easy as it consumes less energy than the conventional coagulation treatment [57]. However, utilizing inorganic salts like aluminum chloride and aluminum sulfate as the coagulation agent has now become controversial because of their possibility of contributing to Alzheimer's disease [56–58]. Polyacrylamide-based materials are also often utilized in the coagulation process, but the possible release of monomers is now considered damaging due to their entering into the food chain and causing potential health impacts (e.g., carcinogenic effects).

Adsorption removal method is a simple and effective method/design since it is easy to use and can be implemented for dye treatment even in small plants; however, it usually produces huge amounts of sludge, especially in the wastewater with high dye concentrations [59]. Adsorption of dyes on many adsorbents (e.g., SiO_2, Al_2O_3, CaO, MgO, Fe_2O_3, Na_2O, K_2O, bentonite, and montmorillonite) has been broadly studied, but the activated carbon has been proven to be the most effective catalyst due to its high specific surface area, ultra high adsorption capacity, and low selectivity for both nonionic and ionic dyes. However, it has some limitations, including the need for regeneration after exhausting, high cost of the activated carbon, and the lack of adsorption efficiency after regeneration [59, 60]. Taking all these facts into consideration, much of the present work involves the degradation and mineralization of synthetic dyestuff in industry by heterogenous photocatalyst.

The heterogenous photocatalyst relates to the water decontamination processes that are concerned with the oxidation of biorecalcitrant organic compounds [4, 61, 62]. This impressive method relies on the formation of highly reactive chemical species that degrade a number of recalcitrant molecules into biodegradable compounds and is known as the advanced oxidation process (AOP).

The Environmental Protection Agency (EPA) has approved the inclusion of AOP as the best available technology to meet the standard with specifications that provide safe and sufficient pollution control of industrial processes and remediation of contaminated sites [42, 63].

Advanced oxidation processes are based on the production of hydroxyl radicals which oxidize a wide range of organic pollutants including dyes quickly and nonselectively. AOPs include homogenous and heterogeneous photocatalytic oxidation systems. The homogenous photocatalytic oxidation system employs various oxidants such as H_2O, O_3, Fenton reagent, NaOCl, and many others either alone or in combination with light [64] (Figure 2). Recently, heterogeneous photocatalysis has emerged as an important degradation technology leading to the total mineralization of organic pollutants, especially synthetic dyes [5, 37, 38, 65, 66].

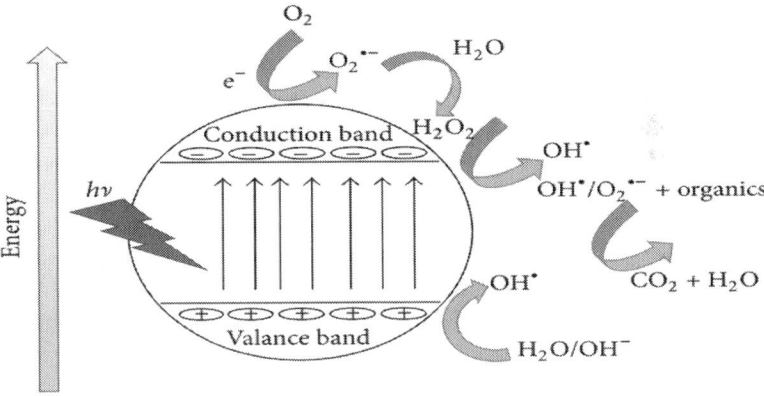

Figure 2: General view on photocatalytic mechanism and degradation process.

Photocatalytic Decolorization of Acid Dyes-Water Soluble Anionics

Acid dyes are chemically a sodium (less often ammonium) salt of a carboxylic or phenol organic acid, or sulfuric acid, with ionic substitution to be soluble in water and contains affinity for amphoteric fibers, while lacking direct dye affinity for cellulose fibers (via hydrogen bonding, Van de Waals, and ionic bonding) [67, 68]. Acid dyes consist of several compounds from the most varied categories of dyes, which represent characteristic differences in structure (e.g., nitro dyes, triphenylmethane, and anthraquinone) [69]. Acid dyes are commonly divided into several classes which depend on level dyeing properties, fastness requirements, and economy, which are indicated by the strength of the anionic characteristic of dyes to the cationic sites of the cellulose fibers [68]. Most of acid dyes are generated from chemical intermediates, where anthraquinone-like structures and triphenylmethane predominate as the final state, which give blue, yellow, and green color [68–70].

Acid dyes, just as any of the synthetic dyes, have the capability of persuading sensitization in humans because of their complex molecular structure and the way in which they are metabolized in the body. Moreover, their water solubility is harmful to human beings since they are sulphonic acids [71]. The sulphonate groups are spread evenly along the molecule on the opposite side to the hydrogen bonding – OH groups, to minimize any repulsive effect [69]. This in consequence determines the main problem with anionic dyes, which is the lack of fastness during the washing and removing process.

Thus, many research works have paid increasing attention to the degradation of acid dyes in the water stream in recent years. Several techniques, including the use of activated carbon, membrane filteration, adsorption, and coagulation have been known to unravel the problems caused by the presence of acid dyes (Table 2).

Table 2: Types of adsorbents used with different anionic/acid dyes

Adsorbent	Anionic dyes	References
Organo-bentonite	Acid scarlet	[72]
	Acid turquoise blue	[73]
	Indigo carmine	[74]
Ammonium functionalized mesoporous materials	Reactive brilliant red	[75]
	Acid fuchsine	[76]
	Orange IV	[77]
	Methyl orange (MO)	[78]
Apatitic tricalcium phosphate	Reactive yellow 4	[79]
Apatitic octocalcium phosphate		[80]

Wood shaving bottom ash	Red reactive 141	[81]
Bagasse ash	Acid blue 80	[82]

However, due to the recalcitrant nature of acid dyes and the high salinity of wastewater containing acid dyes, these conventional treatment processes are feckless. Adsorption and coagulation methods have also been applied to treat acid dyes in wastewater, which always result in secondary pollutants [66]. Furthermore, it is noted that acid dyes have $-SO_3^-$, $-COOH$, $-OH$, and hydrophilic groups and excellent solubility in the water stream [74, 75]. Their molecules spread linearly in solution and have a notable tendency to aggregate by hydrogenous bonding, and consequently form colloids in solution and also tend to be adsorbed and flocculated [81]. To overcome such limitations, photocatalytic decolarization of acid dyes water soluble is essential. This process done through the formation of electron-hole pairs with proper photon energy. It has been assumed that once the energy is larger than the band gap, the electron-hole pairs are separated between the semiconductor's valence and conduction bands [61, 82]. The acid dyes as adsorbed species on suitable sites on the surface of semiconductors undergo photooxidation, reduction, and synthesis under either ultraviolet, sunlight, or even ultrasonic lights. In addition, the aromatic linkages are susceptible to reduction under light irradiation [83] (Figure 3).

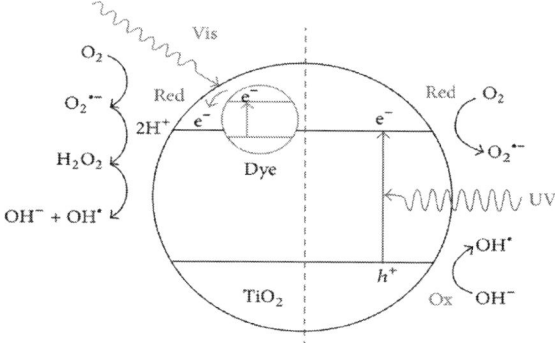

Figure 3: The photocatalytic decolorization of TiO_2 towards Acid Red 44 as a model of acid dyes [83].

This encourages a promising technology based on the advanced oxidation process that has been studied extensively through a broad range of acid dyes that can be nonselectively oxidized quickly [43, 84, 85]. Photocatalysis of acid dyes entails the formation of adequate concentrations of highly reactive transitory species like hydrogen peroxide, hydroxyl radicals, and superoxides to react with acid dyes and degrade them in the presence of a semiconductor and visible light or ultraviolet (UV) light [86]. Usually the semiconductors with band gap energy of 3.2 eV are used as photocatalysts with the assumption that as a proton at equal or higher energy (λ < 400 nm) illuminates the semiconductor, the photon energy creates an electron to jump from the valence band to the conduction bands, generating electrons and positively charged holes [51, 87]. These electron-hole pairs persuade a series of reactions, which oxidize the dye acids.

Among the various semiconductor oxides, TiO_2 and ZnO have been intensively investigated since the discovery of their ability to photocatalyse acid dyes [37]. Briefly, once the aqueous semiconductor (TiO_2 and/or ZnO) suspensions are irradiated in light energy greater than the band gap energy of the semiconductors, conduction band electrons and valence band holes are generated [51, 88, 89]. As the charge separation is maintained, the electrons and holes may migrate to the semiconductor surface where it takes part in the redox reaction with acid dyes [90–92]. The photogenerated electrons react with the adsorbed acid dye molecules (O_2) on the semiconductor site and diminish it to superoxide radical anion ($O_2^{\bullet}$) while the photogenerated holes oxidize the H_2O or OH^- ions adsorbed at the semiconductor surface to $OH^{\bullet}$ radicals [43,93–95]. These generated radicals with other highly oxidant species act as strong oxidizing agents which could easily attack the adsorbed acid dye molecules or those located close to the surface of the semiconductor, thus resulting in complete degradation of acid dyes into its smaller biodegradable fragments [89, 96].

Despite the many benefits of using TiO_2 and ZnO as a photocatalyst to degrade the dye acids, if the aim is to expand a solar-powered treatment technology, there are few disadvantages of the technology that barricade commercialization. Even if both semiconductors offer high absorption and surface areas, they can be adjusted by preparation parameters [84, 97, 98]. Although many acid dyes can be effectively photodecomposed using TiO_2 and/or ZnO as the photocatalyst, the kinetics and mechanism of photocatalytic decolorization with respect to both semiconductors as photocatalysts are comparatively unclear. It has been recorded that both semiconductors can contribute to the decomposition reaction in different ways without decreasing their activity over time [99]. Several kinetic models for catalyzed oxidation utilizing heterogenous catalyst supported by both organic and inorganic carriers have been published in the literature [51, 83, 100]. However, only a few kinetic models of catalyzed photocatalytic decolorization of acid dyes were published. The Mars-Van Krevelen mechanism stated that the surface of the semiconductor catalyst acted as redox mediator, which transferred electrons to oxygen to form oxygen anions as radicals, $O_2^{-\bullet}$. The $O_2^{-\bullet}$ anion radical oxidized the adsorbed acid dye compounds to form various products, while the reduced form of O_2^{-} could be regenerated by gaseous oxygen [61, 101]. The stationary-state adsorption mechanism was based on the steady-state assumption and also the oxidation reduction of the adsorbed phase [102]. The Ely-Rideal mechanism envisaged that a heterogeneous reaction took place among strongly chemisorbed acid dye atoms and physically adsorbed molecules which become attached to the surface by faint Van der Waals forces [84]. The Langmuir-Hinshelwood mechanism is based on the reaction that occurred between both acid dyes and semiconductors [95, 103].

Photocatalytic Decolorization of Basic Dyes-Water Soluble Cationics

Water soluble basic dyes are commonly considered as the most difficult to eliminate or degrade from the dyeing effluent, because of their high stability and resistance ability in the water stream [104–106]. Basic dyes possess cationic functional groups such as $-NH^{3+}$ or $=NR^{2+}$ [105]. Both of these protein functional groups in basic conditions generate a negative charge as the $-COOH$ groups are deprotonated to give $-COO^{-}$ [107]. Basic dyes perform weakly on natural fibers but work very well in acrylics [105]. Basic dyes will form a covalent bond with the proper polyacrylic functionality, and once attached, these basic dyes are very difficult to remove [106]. Cationic triphenylmethane dyes are one of the most extensive basic dyes utilized as colorants and antimicrobial agents in different industries. Previous articles demonstrate that it may further serve as targetable sensitizers in photodestruction of specific cellular components or cells [107, 108]. Methyl green (MG) is a basic triphenylmethane and dicationic dye frequently utilized for staining of solutions in biology and medicine. It is also utilized as a photochromophore to sensitize gelatinous films [109]. The increasing interest in the development of modern and new methodologies for the degradation of toxic basic dyes has led to the deduction that the most effective way for oxidation of the basic dyes is with a powerful oxidizing agent, specifically when a free radical like $^\bullet OH$ is generated [110–112] (Figure 4).

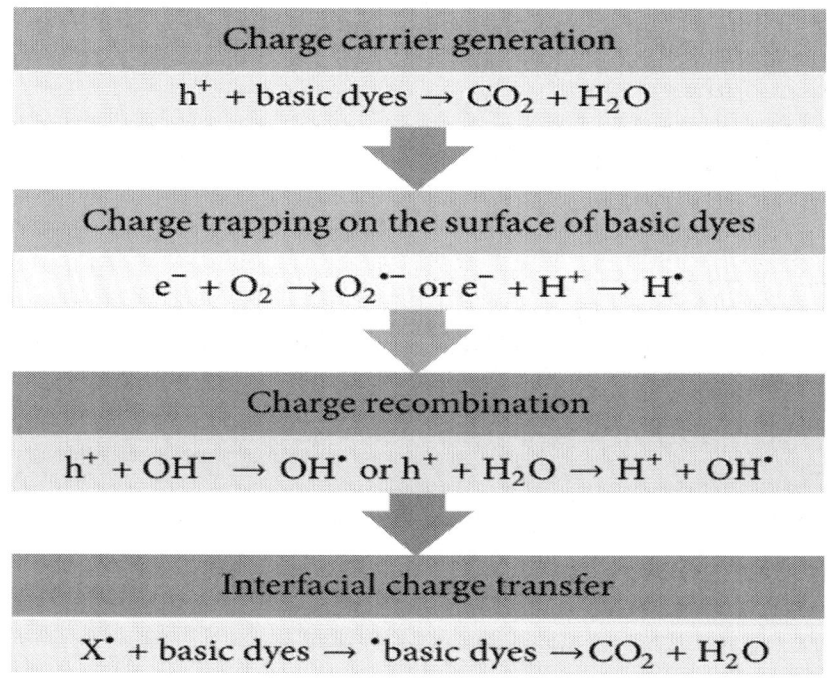

Figure 4: The steps in the photocatalytic process of basic dyes using TiO_2 or ZnO.

Lately, advanced oxidation processes have been broadly investigated and have become alternative methods for decolorizing and reducing recalcitrant wastewaters generated by basic dyes. Likewise, the use of cadmium oxide (CdO) nanostructure as one of the promising semiconductors for this operation demonstrates positive results [113–115]. CdO is an n-type semiconductor with a direct band gap of 2.2 to 2.5 eV and an indirect band gap of 1.36 to 1.98 eV [114]. Since CdO has a band gap tailored to the visible region of solar light with a similar photocatalytic mechanism to semiconductor oxides, it can be an important option as photocatalyst materials especially in the decolorization process of basic dyes [45, 113]. Indeed, the evaluation of photocatalytic activity of CdO towards basic dyes is considered as cauliflower-like [116]. The nanostructure of CdO for removing the basic dyes from aqueous solution has been reported and it is believed that the crystal orientation, morphology, crystallinity, particle size, architecture types, and oxygen defects play an important role in changing the band gap. Actually, diversity in the band gap energy is highlighted to lattice defects because of the Burstein-Moss effect. Besides, the catalytic, optical, and electrical properties originate from the difference of band gaps in different structures [115]. Thus, it is critical to probe an investigation on the generation of new CdO structures for better photodegradation of basic dyes. Different structures of CdO on a nanoscale have been reported, such as nanowires, nanoparticles, nanoneedles, thin film, nanocrystal, and others [117]. CdO micro- and nanoarchitectures with three-dimensional structures such as rods, tubes, and cauliflower-like structures have a larger specific surface area and enhanced oxygen vacancy, which in turn increases the degree of oxidation process on basic dyes [118]. Cauliflower-like architectures have attracted great interest due to its special and novel morphology with high specific surface area that can facilitate the diffusion and mass transportation of the basic dye molecules in photodegradation applications [116]. This particular structure can be easily synthesized using mechanochemical methods, a cheap process, followed by thermal treatment conforming to the detailed process presented in former studies.

Most studies related to photodegradation techniques have been done using TiO_2 and/or ZnO as the model photocatalyst because of their nontoxicity, cheapness, chemical stability, and high photocatalytic activity [37, 119–121]. The photocatalytic decolorization of basic dyes with TiO_2 and/or ZnO as the charge carrier or generation is summarized in Figure 2. The OH$^\bullet$ or the directly produced charge is a strong oxidizing agent which attacks basic dyes present at or near the surface of the semiconductor [122]. It ultimately causes the complete degeneration of the basic dyes into harmless compounds. In general, two different types of TiO_2 phase are normally used in photocatalytic decolorization of basic dyes: anatase (3.2 eV) and rutile (3.0 eV). The adsorptive affinity of anatase for the basic dyes is higher than that of rutile, and thus anatase is generally regarded as the more photocatalytic active phase of TiO_2, presumably due to the combined effect of lower rates of recombination and higher surface sites [123, 124].

The dye derivative reactive brilliant blue (KN-R) has been broadly utilized as a model of basic dyes in the photocatalysis process. The effects of key operational factors like reaction pH, catalyst loading, H_2O_2 dosage, and the initial basic dye concentration on the decolorization were extensively studied to optimize the process for maximum degradation of basic dyes [125]. It can be concluded that the photocatalytic decolorization process performed a fast oxidation without the formation of polycyclic products and intermediate products at a suitable wavelength of light [51, 126]. The reactions frequently take place on the surface of the semiconductors. Hence, the need for a semiconductor supported by a good adsorbent is much felt because of the power to concentrate pollutants near semiconductor particles and the capacity for adsorption of generated intermediates and the capability of reusing adsorbents [127]. In addition, to ensure full use of the solar energy source, it is of great interest to develop photocatalysis of basic dyes for expansion of the adsorption to the visible light range. For both TiO_2 and/or ZnO, a great deal of effort has been focused to extend their photoadsorption to the visible light range, for example, by doping with anions of C, S, and N or transition metal cations [128, 129]. Besides TiO_2 and/or ZnO, a great deal of attention has also been focused in the search for semiconductor oxides of Bi_2WO_6, $BiM_{ox}O_6$, Bi_2, $M_{ox}W_{1-x}$ and $Bi_4Ti_3O_{12}$ which have been recently revealed to exhibit photocatalytic activity and decolorization of basic dyes in the visible light range owing to their lower band gap than that of TiO_2 and/or ZnO [130–140].

Photocatalytic Decolorization of Disperse Dyes—Alkaline

Disperse dyes have low solubility in water. However, they can interact with the polyester chains by forming dispersed particles. Their main application is the dyeing of polyester, and they find less use in dyeing cellulose acetates and polyamides [141–145]. The general structure of disperse dyes is planar, small, and nonionic, with attached polar functional groups such as $-NO_2$ and $-CN$. In addition, this type of dyes is a mitotic toxication agent and should be considered as a biohazard component [142]. Thus, discharge of disperse dyes have become a subject of concern in the universe due to its harmful and toxic effects to living organisms and the environment [143]. As far as the wastewater treatment technologies are concerned, different techniques have been utilized for the reduction and degradation of dispersed dyes such as chemical precipitation, H_2O_2 adsorption, oxidation by chlorine, electrochemical treatment, ozone electrolysis, adsorption, ion pair extraction, flocculation, coagulation, membrane filtration, and specially the photocatalytic process [145–147].

The dispersed dyes (alkaline compounds) can be most effectively decomposed by photocatalytic methods [148–150]. Recently, owing to their unique and special electrical and optical properties, semiconductor materials have gained global acceptance for alkaline dispersed dye treatments [151]. It has been demonstrated that the photooxidation of CN to OCN occured during the photodegradation of alkaline dyes in the presence of powerful oxidation agents [152–154]. Considering that disperse alkaline dyes cannot be treated by conventional biological processes, intensive investigations on the latest treatment techniques of these wastewaters have been conducted to develop effective methods for the remediation and treatment of a wide variety of alkaline-dye pollutants owing to their capability to produce a complete degradation process. The photocatalytic degradation reaction is usually conducted for compounds dissolved in water-like alkaline dyes, at mild temperature and pressure conditions, utilizing ultraviolet-illuminated semiconductor powders without the requirements of expensive oxidants [86, 155–158] (Figure 5).

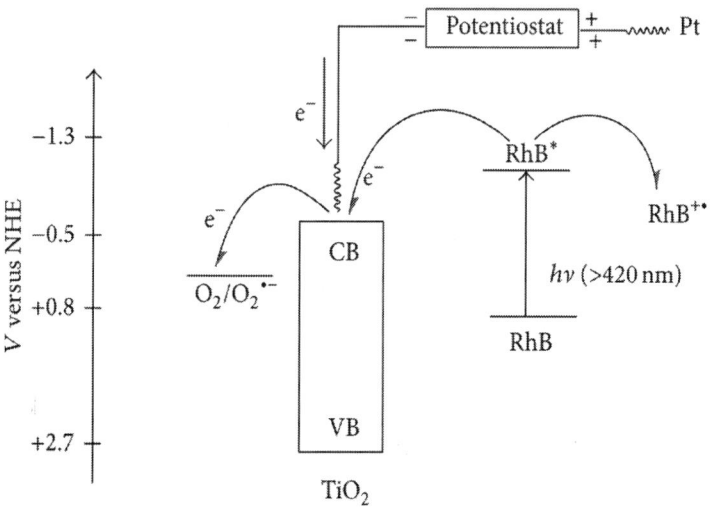

Figure 5: Proposed mechanism of the photoelectrocatalytic degradation of Rhodamine B with TiO$_2$ as the electrode [155].

A semiconductor is generally characterized by the band gap energy between its electronically populated valence band and its broadly vacant conduction band [33].

Copper oxide (CuO) is a one of the most promising semiconductors used in advanced oxidation processes for degradation of alkaline dyes [159–161]. With an energy band gap of 1.21 to 1.5 eV it has the ability to perform under irradiation in sunlight. Reactions involving Cu$^+$/Cu^{2+} lead to the oxidative transformation of alkaline dyes. The unique electronic structure of Cu allows for the interaction with the spin restricted O$_2$ enabling Cu to participate in the redox reaction with alkaline dyes [162]. Many researchers have anticipated the reaction of CuO on different adsorbents like activated alumina, zeolite, or activated carbon in wastewater treatments. It was found that in order to achieve an efficient, stable, and economical catalyst, CuO semiconductors must be fixed on an ideal and an inert support [163–165]. Among all CuO supported systems for alkaline dye photodecolorization, zeolite was found to be the most ideal with several distinct advantages, including super adsorption capability, unique uniform pores, and special ion-exchange capability [166].

Photocatalytic Decolorization of Vat Dyes-Water Soluble Alkali Metal Salts

Almost 22% of the total volume of industrial wastewater produced comes from the textile industry, with 7×10^5 tonne of materials classified as vat dyes or water soluble alkali metal salts [167, 168]. This type of dyes produces undesirable effluents and is discharged into the environment without further treatment. Once the vat dyes enter natural water bodies, it can cause intense problems if not treated, since the dyes are toxic, mutagenic, and carcinogenic to human life as well as can inhibit photosynthesis of aquatic life even in quantities as low as 1 ppm [169]. To solve this problem, several semiconductors for oxidative photodecolorization have been tested, including WO_3, TiO_2, MnO, CuS, ZnO, Fe_2O_3, ZrO_2, CuO, CdS, ZnS, In_2O_3, SnO_2, and Nb_2O_5 [170–176].

The selection of the type of semiconductor is based on its ability to convert the vat dyes into nontoxic products [44, 90, 177, 178]. Additionally, the use of mesoporous materials, like zeolite as a support for these series of semiconductors, has recently become the focus of intensive research on vat dye photodecolorization, due to the fact that the semiconductor support influences the photocatalytic efficiency through structural features, and the interaction between the vat dyes leads to the enhancement of contact between the surface, and irradiation likewise decreases with the amount of semiconductor required [179]. Thus, there are some studies focused on the importance of semiconductor supported zeolite for vat dye photodegradation, including Co-ZSM-5, TiO_2-HZSM-5, Fe-exchange zeolite, and CuO-X zeolite [180, 181].

TiO_2 has been the most studied material for the photocatalysis of vat dyes [182–186]. ZnO has also been identified to be the main contender whose physicochemical properties are comparable to those of TiO_2. However, ZnO, just as with TiO_2, suffers from its large band gap energy that is close to 3.2 eV, which limits its adsorption of solar light emission that reaches the earth to less than 3 to 4%. For both semiconductor materials, the valence band is composed of O^{2-} (2p orbital), which is of anionic character, that induces interactions and oxidation composition with the vat dyes (alkali metal salts) [187–190]. However, this anionic character gains rapid recombination and holes during the oxidation process and in turn diminishes the efficiency of the photocatalytic reaction [191, 192]. From this viewpoint, efforts have been allocated to extend the adsorption of TiO$_2$ and ZnO to deal with the photosensitization by vat dye molecules. According to the literature, reports on ZnO for the vat dye photodegradation are still scarce. The nanosized ZnO has extracted intense interest in recent years, especially in vat dye photodegradation for enhancing its performance, such as changes in surface properties and increase in surface area as well as in quantum effects of the overall decolorization process [193–195]. Upon light irradiation, this nanosized ZnO produced a highly active radical species that can quickly oxidize the vat dyes into harmful residues.

Copper sulfide (CuS) is another type of semiconductor that is currently used in photocatalytic decolorization of vat dyes [196–199]. CuS with a layered structure is a transparent p-type semiconductor with a band gap above 3.1 eV [198]. The top of the valence band is principally composed of well-hybrid state of Cu 3d and S-3p states, while the bottom of the conduction band consists primarily of Cu 4s state [196, 197]. The band gap of CuS was recognized to be a direct-allowed transition type through the analysis of the symmetry [199]. This small dispersion of the conduction band leads to the broadband gap and high stability of CuS to be more convenient for vat dye adsorption and oxidation.

Photocatalytic Decolorization of Azoic Dyes

The azoic dyes that are normally used on industrial scale have characteristics that are dependent on one or more azo bonds (–N=N–) with aromatic rings [167, 200–202]. The aromatic ring system of the dyes helps to strengthen the Van-der-Waals forces between dye and fibers [203]. Most synthetic dyes have significant structural variations and are extremely stable in performance under light and washing and most severely are resistant to aerobic biodecolorization by bacteria [204, 205]. Thus, it has been reported that the effluents from textile or dye industries involve aromatic compounds which are chemically stable and harmful to human health [201]. Various substitutions on the aromatic nucleus and most versatile groups of compounds give structurally diverse which make them recalcitrant, xenobiotic, noticeable in public, teratogenic, and resistant to degradation [204]. The amount of azo dye concentrations present in wastewater varied from very low to high concentrations (5 to 1500 mgL^{-1}) that leads to color dye effluents causing toxicity, including carcinogenic and mutagenic effects in biological ecosystems. In addition, under anaerobic conditions acid dyes are promptly decreased to potentially hazardous aromatic amines [206]. Therefore, water soluble azo dyes even at low concentrations can cause water streams to be highly colored. On the other hand, azo dyes are insoluble in water but may become solubilised by alkali reduction, for instance, by sodium dithionite which is a reducing agent in the presence of sodium hydroxide [207]. Hence, they tend not to contain several functional groups which may be assailable to oxidation and reduction, which in turn gives harmful effects to the water stream habitat.

The biological approach of the decolorization of azo dyes takes place either by adsorption on the microbial biomass such as fungi, algae, yeast, and bacteria, along with anaerobic to aerobic treatments or biodegradation by the cell [208]. Azo dyes can also be reduced chemically by sulfide and dithionite. The decolorization mechanism of azo dyes based on the extracellular chemical reduction with sulfide was postulated for sulfate reducing environments [209]. However, it has also been noted that for the treatment of azo dyes containing wastewater, traditional methods like flocculation, adsorption onto activated carbon, activated sludge process, and reverse osmosis have difficulties in complete degradation of pollutants and also have the further disadvantage of resulting in secondary pollution [208–210]. Moreover, anaerobic decolorization of azo dyes may also produce carcinogenic aromatic amines.

Therefore, the photocatalytic oxidation technique has received significant attention for destroying of azo dyes in recent years. This technique can be divided into homogenous and heterogeneous subgroups, based on the action of OH$^{\bullet}$ which enables almost complete mineralization of azo dyes under mild experimental conditions due to the high oxidation potential [211–215]. In heterogeneous photocatalysis of azo dyes, the electron-hole pairs will be initially produced by irradiation of a semiconductor with a photon of energy equivalent to or greater than its band gap width [213]. The electrons and holes may migrate to the semiconductors on the catalyst surface where they take part in redox reactions with the adsorbed azo dyes [212]. The oxidizing radicals could attack the azo dye molecule and disintegrate it into CO_2 and H_2O molecules which are nontoxic [212–214]. It has been suggested that the formation of free radicals acts as a primary oxidizing species [216]. The mechanism on photodecolorization of azo dyes with methyl red and methyl orange as a model of compound is illustrated in Figures 6 and 7, respectively.

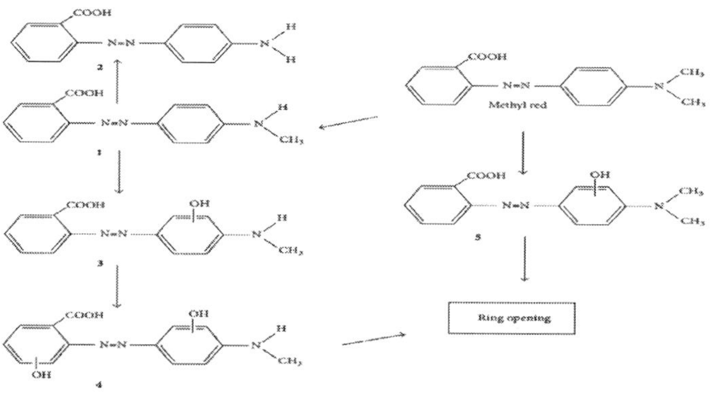

Figure 6: Proposed pathway for the photodecolorization of methyl red [216].

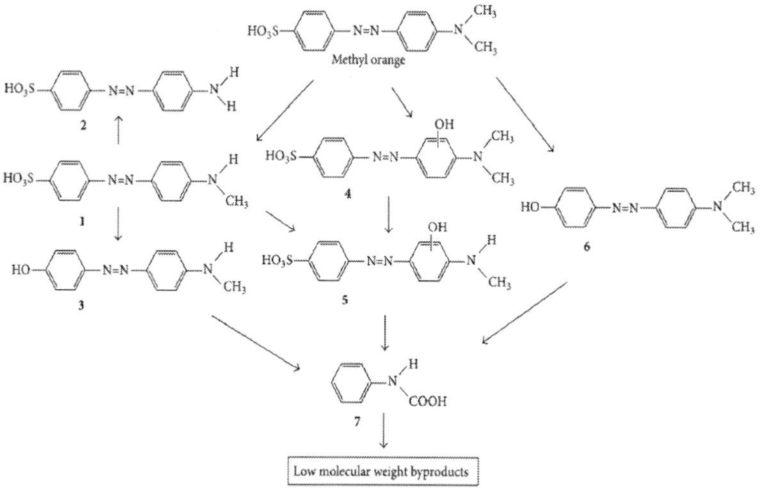

Figure 7: Proposed pathway for the photodecolorization degradation of methyl orange [216].

It is claimed that azo dyes are noted for their photocatalytic decolorization in the absence of oxygen whenever a suitable electron donor or hydrogen source is present [217]. Structurally, azo dyes are double bonded belonging to different chromophoric groups and are heterocyclic and adsorb visible light [208]. The reduction of the chromophoric group shifts the visible region of the ultraviolet or infrared region, and thus a reduction in color is achieved. Consequently this phenomenon has encouraged several research works using heterogenous semiconductor photocatalysts like TiO_2, ZrO_2, SnO_2, Fe_2O_3, CuO, ZnO, and CdSas as an alternative to conventional methods for the degradation of azo dyes from wastewater streams [218–220]. The degradation of hydrophobic and hydrophilic azo dyes has been demonstrated to be effective in acetone solution under exposure to UV light.

Recently, photocatalysis of azo dyes using solar or artificial light and TiO_2 has been the objective of several studies as it is an attractive low energy strategy that has been applied to many other organic compounds (e.g., phenol) [4]. TiO_2 is chemically inert, corrosion resistant, and most importantly, it works under mild conditions without any chemical additives [221,222]. Meanwhile, it was found that in degradation of methyl orange or 4-4-[(dimethylamone)phenylazo] benzenesulfonic acid, a TiO_2 film was up to 50% less effective than the TiO_2 slurry. However, some improvements were observed after coating/doping the TiO_2 film with metals, but the films were still not as impressive as the slurry [223, 224]. Meanwhile, other studies have shown that only cationic azo dyes can be adsorbed on the surface of the photocatalyst and simultaneously their photocatalytic degradation was quicker than the degradation of anionic azo dyes like Eriochrome Black T [225, 226]. It has been found that TiO_2 adsorbed almost only cationic azo dyes, except for the anionic Quinizarin with an adsorption efficiency of 21.8% [227, 228]. Apart from photooxidation, the photoreduction of azo dyes is also known as a significant decolorizing or a folding pathway. This fact can be explained in relation to the surface structure of TiO_2. In the unmodified surface structure of crystal TiO_2, oxygen atoms are mainly present with a high electron density which creates a negative center [228–230]. Thus, the TiO_2 particles have a negative charge and are more suitable to adsorb cationic azo dyes than the ones with anionic characteristics [182, 231, 232]. Furthermore, the modelling of photodecoloration of nonbiodegradable azo dyes was investigated recently with Reactive Red 2 in a cocktail mixture of triethylamine and acetone. It was found that the cocktail photolysis system was able to entirely decolorize the azo dye in a short time and the overall dye removal followed pseudo-first-order decay kinetics [233].

Furthermore, a TiO_2-based photocatalysis for azo dye degradation has been developed. It can be applied as a film and has the effectiveness of the slurry [234–236]. An approach to enhance the photocatalytic reaction rate is by modifying the semiconductor with transition metal. Decorating TiO_2 with other metal/nonmetal or metal/metal combinations can decrease its band gap and allow for activation by the longer wavelength of visible light [237–239]. Hence, solar energy can be used more effectively in the photocatalysis process. Currently, many metals (e.g., Fe, Cu, Co, Al, Cr, Ce, Ag, and Nd) and nonmetals (N, C, F, S, and B) have been attached onto TiO_2 for azo dye degradation [237–242]. Among the metals, Ag^+ has been recognized to be more effective than Fe^{3+}, Co^{2+}, Ce^{4+}, and Cu^{2+}, since it traps the photogenerated electrons and avoids the recombination of electrons and holes [243].

ZnO has been demonstrated to have a much higher efficiency than TiO_2 in the case of azo dyes degradation irradiated by UV light; however, studies on heterojunction systems applied to water treatment have primarily been restricted to the sensitization of TiO_2 [244, 245]. This statement has been supported by the fact that ZnO has numerous advantages over TiO_2. This includes high efficient photocatalytic activity, and photodegradation of diluted azo dyes cannot proceed sufficiently because of insufficient contact between azo dyes and semiconductors. This is an important factor in hindering photocatalytic activity [246]. The mass transfer from azo dyes to the semiconductor surface limits the photodegradation rate of diluted azo dyes. It is important that visible light degradation of some dyes utilizing ZnO was shown to be more effective than TiO_2. In this case the degradation mechanism was based on electron injection from the exited dyes to the ZnO conduction band. This was much more significant as compared to TiO_2 which indicates high efficiency of charge transport and limited charge loss [247–249].

Photocatalytic Decolorization of Sulfur Dyes

Textile industries generate large amounts of colored sulfur dye effluents which are toxic and induce a lot of damage to the environment. In view of the mutagenic character or carcinogenic nature of sulfur dyes, the deleterious effects of the color in receiving water, and the customary resistance of the sulfur dyes to biological degradation, the necessity of investigating new alternatives for appropriate treatment of this kind of dyes is evident [250–252]. Thus, various methods for the removal of sulfur dyes have been reported, including biological and chemical flocculation, coagulation, adsorption and oxidation, electrochemical oxidation, membrane separation, and ion exchange methods [253–255]. These methods have their own limitations for the removal of sulfur dyes, including being expensive, time consuming, and commercially unattractive as well as resulting in the production of secondary wastes [254]. Furthermore, these processes are also ineffective for sulfur dye removal since sulfur dyestuff is biorecalcitrant. In addition, these series of physicochemical treatments prepare only a phase transfer of sulfur dyes and produce huge quantities of sludge [256, 257].

The efficiency of a photocatalytic decolorization reaction is determined by the properties and quality of the photocatalyst, which is often a semiconductor with the ability to create electron-hole pairs under photoillumination [258, 259]. Thus, it is an important step to recognize an efficient and suitable photocatalyst during the decomposition process. In recent decades, different mixed metal oxides consisting of TiO_6, TaO_6, or NbO_6 octahedral units, such as $BaTi_4O_9$, $SrTiO_3$, $K_4Nb_6O_{17}$, $InTaO_4$, and Ni_xTaO_4 had been extensively investigated as a new class of photocatalysts in the field of sulfur dye degradation [260–262]. These kinds of photocatalysts belong to a family of uniform heterogenous catalysts [261]. Yet only a few of these photocatalysts have been studied for the removal of environmental contaminants, and earlier authors have all used the solid-solid blending method to synthesize their sample. Recently, the typical photocatalysts developed are mostly oxides containing d-block element ions as Ti^{4+}, Ta^{5+}, Nb^{5+}, and Zr^{4+} with d0 electron configuration [263]. Very recently, researches have also focused on p-block metal oxide photocatalyst with d10 electron configuration due to their fair mobility for sulfur dye degradation [261, 262]. TiO_2 was found to be the most efficient photocatalyst for photodegradation of sulfur dyes because of faster electron transfer of molecular oxygen [264–266]. Furthermore, TiO_2 photocatalyst is largely available as a nontoxic, inexpensive, and with relatively high chemical stability [42]. It has been noted that the photocatalytic degradation of sulfur dyes in solution is initiated by photoexcitation of the semiconductor, followed by formation of an electron-hole pair on the semiconductor surface [267]. The high oxidation potential of the hole in the semiconductor permits the direct oxidation of sulfur dyes into reactive intermediates [268, 269]. Highly reactive OH• can also be formed either by decomposition of water or by the reaction of the hole with OH⁻. The OH• radical is a very strong, nonselective oxidant that leads to the degradation of organic chemicals [270].

There are certain relationships between properties of dyes and treatment mechanisms. Sulfur dyes are often made of azo compounds, sulfide structures, or anthraquinones, and they have several –C=O, –NH–, and aromatic groups. These dyes tend to be adsorbed by Fe(OH)$_x$ particles [271–273]. However, the photodegradation of sulfur dyes utilizing semiconductors is not new. The sulfur dye treatment of photocatalyst would be more suitable if the semiconductor was immobilized, so the semiconductors would not have to be separated from the sulfur dye solution [274, 275]. Thin films are one of the most important technological applications. Thin film photocatalyst towards sulfur dyes photodegradation offers high stability and convenient reuse and hence has received more and more attention [276]. Furthermore, photocatalysis supports such as zeolite have been extensively used to enhance the photodegradation of sulfur dyes. Zeolites are crystalline aluminosilicates with cavities in which the size can change in the range from one to several tens of nanometers depending on the type of aluminosilicate framework, Al/Si ratio, and the origin of the ion exchange cations [277–279]. These characteristics of zeolite make it more selective for photocatalytic oxidation and are crucial especially when using environmentally benign oxidants.

RECENT ADVANCES IN SYNTHETIC DYES PHOTOCATALYTIC DECOLORIZATION

Industrial effluent detoxification is one of the most challenging global problems. Dyes, phenols, pesticides, fertilizers, detergents, herbicides, surfactants, and other synthetic organic compounds are disposed of directly into the environment, without being treated, controlled, or uncontrolled, without an effective treatment strategy [280–282]. Their toxicity, stability to natural decomposition, and persistence in the environment have been the cause of much concern to societies and regulatory authorities around the world [283, 284].

Although the strong potential of photocatalytic process for wastewater treatment is widely recognized via numerous patents and publications, technical development at industrial level has not been met with much success [285–287]. This is due to the high operating cost of the photocatalytic oxidation process relative to existing biological treatments [288]. Since in tropical countries, sunshine is available in abundance; therefore, application of this oxidation technology using solar light can be a cost- and energy-effective detoxification technology. Furthermore, the limitations of the photocatalyst system can be addressed in terms of the tight range of pH in which the reaction proceeds, the requirement for recovering the precipitated catalyst after treatment, and the deactivation by some ion-complexing agents such as phosphate anions [289, 290].

Using solar energy is an interesting aspect in photocatalyst technologies. Solar photocatalysis has become an important area of research in which sunlight is the source of illumination to perform various photocatalytic reactions with regard to different kinds of dyes [291, 292]. As visible light is the main component of solar radiation, the development of a stable photocatalytic system, which can be affected by visible light, is most probably indispensable. In order to overcome the limitations, many studies on coupled semiconductor photocatalysts like ZnO-TiO_2, CuO-ZnO, CuO-TiO_2, CuO-SnO_2, TiO_2-SnO_2, ZnO-SnO_2 and so on have been reported [198, 293–295]. These series of binary oxide photocatalysts showed enhanced catalytic activities and selectivities compared to the monocomponent photocatalyst. This combined system also provides a more controllable rate of recombination as the composition of two semiconductors with different band gaps can suppress the recombination of e^-/h^+ pairs [296]. Amongst the series of binary systems, CdS/TiO_2 showed the most prospect as an effective visible light photocatalyst for dye reduction and degradation. In the system of TiO_2/CdS, the photogenerated electrons in CdS are transferred into the TiO_2 particle, while the holes remain in the CdS particles [297, 298]. This combination has also overcome the limitation of native CdS as photocatalyst due to its photocorrosion. Other researchers have loaded semiconductor with carbon-based nanomaterials like activated carbon, CNTs, graphene, graphite, and other matrices to improve the photocatalytic activity or cycling and its ratings performance [299–302]. Meanwhile, recent research has indicated that organic polymer films such as chitosan and cellulose films can ensure the stabilization of semiconductors especially in nanosized form and also provide an interface for the charge transfer and correspondingly improve photocatalytic efficiency [301–303]. In addition, the incorporation of such biopolymers assist in reducing the leakage of semiconductor particles in treated water during the dye removal and degradation, since those types of biopolymers are effective adsorbents and chelators for semiconductor ions in aqueous solutions [304].

INFLUENCE OF DYE TYPE ON THE PHOTOCATALYTIC PROCESS

The chemical structure of the organic dyes has a considerable effect on the reactivity of dyes on photodegradation system [301]. This effect has been explored by different researchers. For example, the COD removal rate of RY17 was found to be higher than RR2 and RB4 dyes. This is due to the structural difference among the three molecules of dyes. RY17 and RR2 are equipped with an azo group ($-N=N-$), which is not present in RB4 molecules and suspected to photodegradation. In addition, $-CH_2-OS_2-$ linkage in RY17 is also labile in the reaction environment. In RB4, the presence of anthraquinone structure and the absence of azo band make it resistant to photodegradation [305].

Meanwhile, the removal of reactive orange 16 was maximum, closely followed by reactive blue 4 and reactive 5 in case of TiO_2 photocatalysis. It may be due to the difference in chemical structure of dyes, resulting in difference in adsorption characteristics and difference in susceptibility to photodegradation [43, 44, 47]. The chemical structure of the dyes indicates that reactive black 5 has more complex structure, making it less photodegradable. Another reason may be due to absorption of light photon by dye itself leading to a less availability of photons for hydroxyl radical generation. It was observed from the absorption spectra of three dyes in near UV range that reactive black 5 strongly absorbs near UV radiation compared to reactive orange 16 and reactive blue 4, leading to less by the dye molecules is thought to have an inhibitory effect on the photogeneration of holes or hydroxyl radicals, because of the lack of any direct contact between the photons and immobilized TiO_2. [48]. Indeed, it causes the dye molecules to adsorb light and the photons never reach the photocatalyst surface; thus, the photodegradation efficiency decreases.

It is also important to notice that degradation pathway of organic dyes may be different as according to the chemical structure and functional groups. For example, with an addition of a $^\bullet OH$ radical to an aromatic ring of dyes molecules, a labile H atom is produced [56–60]. This mechanism is also unsatisfactory for hydroxy azo dyes (AO7 and AO8). In that case, abstraction of the H atom, carried by an oxygen atom in the azo form and by a nitrogen atom in the hydrazone form, competes with the addition of $^\bullet OH$ radical on a phenyl or naphthyl nucleus [72].

The functional groups in the chemical structure of dye could be nitrite groups, alkyl side chain, chloro group, carboxylic group, sulfonic substituent, and also hydroxyl groups [305]. The appropriate photocatalyst material has to be chosen depending on these functional groups in the chemical structure of dye [81–84]. Every group that tends to decrease the solubility of molecules in water will decrease the degradation process. In order to evaluate the influence of a nitrite group, the degradation of an analogous pair of dyes such as Acid Red 29 and Chromotrope 2B can be mentioned. Chromotrope 2B contains a nitrite group in the para position with respect to the azo function [305]. This substituent interacts with the phenyl ring and there is a consequent delocalization of the p electrons of the ring and of the unpaired electrons of the heteroatom. As a result, the phenyl ring is electron-enriched, and the nitrite group thus favors attack of an electrophilic entity. The experiment confirms this hypothesis: Chromotrope 2B reaction rate is slightly higher than that of AR29. Hydroxyl radicals have a very short lifetime, so that they can only react where they are formed [72]. Therefore, oxidation reactions can only be successfully performed in homogeneous media. As it was previously mentioned, every group that tends to decrease the solubility of molecules in water will decrease the degradation process. This explains why the rate of decomposition clearly decreases with increasing length of the side chain and consequently with increasing hydrophobicity of the dye molecule, as seen at the degradation of AB25 and RB19 [81–85]. A parallel reaction may take place between •OH radical and hydrogen atoms of the side chains. This reaction competes with destruction of the dye chromophore, without leading to a decrease in the absorbance of the solution.

Considerable decrease of photocatalytic decolorization rate was observed when two or three chloro substituents were present on the phenyl ring of a pyrazolone dye [104, 170]. Indeed, comparison of acid yellow 17 and acid yellow 23 decolorization rates suggests that the difficulty of the dye to be degraded directly depends on the number of electron withdrawing chloro groups in the molecule. The decolorization kinetics of acid yellow 17 is less than those of acid yellow 23.

The photocatalytic decolorization of four organic dyessuch as Alizarin S, Orange G, Methyl Red, and Congo Red by UV/TiO$_2$ has been processed to explore the effect of the presence of carboxylic substituent in dye chemical structure. The photocatalytic rate constants were in the following order: Methyl Red > Orange G ≈ Alizarin S > Congo Red [274]. It has been explained that the higher degradability of MR could be due to the presence of a carboxylic group which can easily react with H$^+$ via a photo-Kolbe reaction. However, the presence

of a withdrawing group such as $-SO_3^-$ is probably at the origin of the less efficient Orange G and Alizarin S degradations [274]. Another suggestion to explain the different reactivity of these dyes could also be their ability to adsorb on TiO$_2$ surface.

Unexpectedly, the presence of the more powerful electron withdrawing sulfonic group on a molecule makes it only very slightly less sensitive to oxidation. Indeed, molecules with one, two, or three sulfonic functions have almost the same reactivity with respect to oxidation by hydroxyl radicals [63, 83, 169, 209]. Acid red 14 containing two sulfonic groups is more reactive in a photocatalytic degradation process in comparison with acid red 18 and acid red 27 that contain three sulfonic substituents [169]. Study of the influence of the sulfonic group is very difficult, because this substituent operates in different fields: it decreases electron density in the aromatic rings and the β nitrogen atom of the azo bond by −I and +M effects. On the other hand, it increases the hydrophilic-lipophilic balance of the dye molecules and consequently slows down their aggregation degree [63, 209].

The electronic properties of a hydroxyl group are −I and +M effects. That is why the photocatalytic decolorization rate of acid red 29, which contains two hydroxyl substituents, is more than that of orange G, which contains one hydroxyl substituent [83]. In both dyes, one molecule contains a hydroxyl group next to the azo bond. But the resonance effect of a substituent operates only when the group is directly connected to the unsaturated system. Therefore, to explain the effect of the hydroxyl group on the reactivity of the organic matter, only the field effect (−I) must be considered. The number of hydroxyl groups in the dye molecule can intensify this resonance and, consequently, the degradation rate of the dye [209].

Photocatalytic decolorization rate of monoazo dyes is higher than dyes with anthraquinone structure. The presence of methyl and chloro groups in the dye molecule decreases slightly the process efficiency while a nitrite group acts in an opposite direction [305]. Alkyl side chain decreases the solubility of molecule in water and consequently disfavors the photocatalytic degradation process. The dyes which contain more sulfonic substituents are less reactive in the photocatalytic process, while hydroxyl group intensifies the electron resonance in the molecule and the degradation rate of the dye. Photocatalytic decolorization takes place at the surface of the catalyst. Dye molecules adsorb onto the surface of photocatalyst material by electrostatic attraction and get mineralized by nonselective hydroxyl radicals. Therefore, the adsorption of the target molecule on photocatalyst material surface may be regarded as a critical step toward efficient photocatalysis.

CONCLUSIONS

In the textile industry, regulations concerning the discharge of wastewater have become more and more stringent. The synthetic dyes utilized in the textile and other industries generate hazardous waste. The dye is utilized to impart color to materials of which it becomes an integral part. However, dye removal is an important but challenging area of wastewater treatment since some dyes and their degradation products are carcinogenic and toxic to mammals. Destructive oxidation of poisonous dyes via photocatalytic approaches have recently received considerable attention since colored aromatic compounds have proven to be degraded effectively by a variety of heterogenous semiconductor catalysts. Photocatalysis aims at mineralization of poisonous dyes to CO_2, H_2O and inorganic compounds or at least their transformation into biodegradable or harmless products. Finally, taking into account that UV light is not only expensive but also harmful to aquatic life, there is the need to improve the ability of photocatalysts to work with visible light.

ACKNOWLEDGMENTS

This work is financially supported by University Malaya Research Grant (UMRG RP022-2012E) and Fundamental Research Grant Scheme (FRGS: FP049-2013B) by Universiti Malaya and Ministry of High Education, Malaysia, respectively.

REFERENCES

1. W. Zhang and C. W. Wu, "Dyeing of multiple types of fabrics with a single reactive azo disperse dye," Chem Papers, vol. 68, pp. 330–335, 2014.

2. S. Nagai, "Induction of the respiration-deficient mutation in yeast by various synthetic dyes," Science, vol. 130, no. 3383, pp. 1188–1189, 1959.

3. A. Heinfling, M. J. Martinez, A. T. Martinez, M. Bergbauer, and U. Szewzyk, "Transformation of industrial dyes by manganese peroxidases from Bjerkandera adusta and Pleurotus eryngii in a manganese-independent reaction," Applied and Environmental Microbiology, vol. 64, no. 8, pp. 2788–2793, 1998.

4. Y.-C. Hsiao, T.-F. Wu, Y.-S. Wang, C.-C. Hu, and C. Huang, "Evaluating the sentizing effect on the photocatalytic degradation of fyes using anatase-TiO_2," Applied Catalysis B, vol. 148, pp. 250–257, 2014.

5. S. Zhang, "Preparation of controlled shape Zn S microcrystals and photocatalytic property," Ceramics International, vol. 40, pp. 4553–4557, 2014.

6.   A. Cavaco-Paulo, J. Morgado, L. Almeida, and D. Kilburn, "Indigo backstaining during cellulase washing," Textile Research Journal, vol. 68, no. 6, pp. 398–401, 1998.

7.   H. Duffner, E. Bach, E. Cleve, and E. Schollmeyer, "New mathematical model for determining time-dependent adsorption and diffusion of dyes into fibers through dye sorption curves in combination shades. Part II: kinetic data from dyeing cotton with a trichrome direct dye system," Textile Research Journal, vol. 70, no. 3, pp. 223–229, 2000.

8.   T. Bechtold and A. Turcanu, "Electrochemical vat dyeing combination of an electrolyzer with a dyeing apparatus," Journal of the Electrochemical Society, vol. 149, no. 1, pp. D7–D14, 2002.

9.   H. E. Liang, S. Zhang, B. Tang, L. Wang, and J. Yang, "Dyeability of polylactide fabric with hydrophobic anthraquinone dyes," Chinese Journal of Chemical Engineering, vol. 17, no. 1, pp. 156–159, 2009

10. S. R. Marder, C. B. Gorman, F. Meyers, et al., "A unified description of linear and nonlinear polarization in organic polymethine dyes," Science, vol. 265, no. 5172, pp. 632–635, 1994.

11. C. S. Tidball, "Intestinal and hepatic transport of cholate and organic dyes," The American Journal of Physiology, vol. 206, pp. 239–242, 1964.

12. W. Tanthapanichakoon, P. Ariyadejwanich, P. Japthong, K. Nakagawa, S. R. Mukai, and H. Tamon, "Adsorption-desorption characteristics of phenol and reactive dyes from aqueous solution on mesoporous activated carbon prepared from waste tires," Water Research, vol. 39, no. 7, pp. 1347–1353, 2005. ·

13. O. Greengauz-Roberts, H. Stoppler, S. Nomura et al., "Saturation labeling with cysteine-reactive cyanine fluorescent dyes provides increased sensitivity for protein expression profiling of laser-microdissected clinical specimens," Proteomics, vol. 5, no. 7, pp. 1746–1757, 2005. ·

14. M. Casetta, V. Koncar, and C. Caze, "Mathematical and modeling of the diffusion coefficient for disperse dyes," Textile Research Journal, vol. 71, no. 4, pp. 357–361, 2001.

15. K. Ryberg, B. Gruvberger, E. Zimerson, et al., "Chemical investigations of disperse dyes in patch test preparations," Contact Dermatitis, vol. 58, no. 4, pp. 199–209, 2008. ·

16. P. Lebaron, N. Parthuisot, and P. Catala, "Comparison of blue nuclei acid dyes for flow cytometric enumeration of bacteria in aquatic systems," Applied and Environmental Microbiology, vol. 64, no. 5, pp. 1725–1730, 1998.

17. N. Daneshvar, D. Salari, and A. R. Khataee, "Photocatalytic degrdataion of azo dye acid red 14 in water on azo as an alternative catalyst of TiO_2," Journal of Photochemistry and Photobiology A, vol. 162, no. 2-3, pp. 317–322, 2004. ·

18. R. Russ, J. Rau, and A. Stolz, "The function of cytoplasmic flavin reductases in the reduction of azo dyes by bacteria," Applied and Environmental Microbiology, vol. 66, no. 4, pp. 1429–1434, 2000·

19. H. Xu, T. M. Heinze, S. Chen, C. E. Cerniglia, and H. Chen, "Anaerobic metabolism of 1-amino-2-naphthol-based azo dyes (Sudan dyes) by human intestinal microflora," Applied and Environmental Microbiology, vol. 73, no. 23, pp. 7759–7762, 2007·

20. R. R. Peterson and J. Weiss, "Staining of the adenohypophysis with acid and basic dyes," Endocrinology, vol. 57, no. 1, pp. 96–108, 1955.

21. L. Peters, K. J. Fenton, M. L. Wolf, and A. Kandel, "Inhibition of the renal tubular excretion of N'-methylnicotinamide (NMN) by small doses of a basic cyanne dye," The Journal of Pharmacology, vol. 113, no. 2, pp. 148–159, 1955.

22. M. Wang, N. Chamberland, L. Breau, et al., "An organic redox electrolyte to rival triiodide/iodide in dye-sensitized solar cells," Nature Chemistry, vol. 2, no. 5, pp. 385–389, 2010. ·

23. L. Camarero, R. Peche, J. M. Merino, and E. Rodríguez, "Photo-assisted oxidation of indigocarmine in an acid medium," Environmental Engineering Science, vol. 20, no. 4, pp. 281–287, 2003.

24. S. Blumel, M. Contzen, M. Lutz, A. Stolz, and H.-J. Knackmuss, "Isolation of a bacterial strain with the ability to utilize the sulfonated azo compound 4-carboxy-4'-sulfoazobenzene as the

sole source of carbon and energy," Applied and Environmental Microbiology, vol. 64, no. 6, pp. 2315–2317, 1998.

25. Y. Hong, M. Xu, J. Guo, Z. Xu, X. Chen, and G. Sun, "Respiration and growth of Shewanella decolorationis S12 with an azo compound as the sole electron acceptor,"Applied and Environmental Microbiology, vol. 73, no. 1, pp. 64–72, 2007.

26. A. Sharma, S. Rani, A. Bansal, and A. Sood, "Effect of mordant combination on silk dyeing with apricot dye," in Natural Dyes: Scope and Challenges, pp. 137–143, Scientific Publishers, 2006.

27. K. Harbinder and K. Namrita, "Eco-friendly finishing and dyeing of jute with direct and mordant dye method," Asian Journal of Home Science, vol. 7, pp. 19–22, 2012.

28. G. R. Cameron and G. Scholar, "The staining of calcium," The Journal of Pathology and Bacteriology, vol. 33, pp. 929–955, 2005.

29. N. Panchuk-Voloshina, R. P. Haugland, J. Bishop-Stewart et al., "Alexa dyes, a series of new fluorescent dyes that yield exceptionally bright, photostable conjugates," Journal of Histochemistry and Cytochemistry, vol. 47, no. 9, pp. 1179–1188, 1999.

30. A. Mathur, Y. Hong, B. K. Kemp, A. A. Barrientos, and J. D. Erusalimsky, "Evaluation of fluorescent dyes for the detection of mitochondrial membrane potential changes in cultured cardiomyocytes," Cardiovascular Research, vol. 46, no. 1, pp. 126–138, 2000. ·

31. H. H. Szeto, P. W. Schiller, K. Zhao, and G. Luo, "Fluorescent dyes alter intracellular targeting and function of cell-penetrating tetrapeptides," The FASEB Journal, vol. 19, no. 1, pp. 118–120, 2005·

32. D. A. Hinckley, P. G. Seybold, and D. P. Borris, "Solvatochromism and thermochromism of rhodamine solutions," Spectrochimica Acta Part A, vol. 42, no. 6, pp. 747–754, 1986.

33. F. H. Hussein, "Chemical properties of treated textile dyeing wastewater," Asian Journal of Chemistry, vol. 25, pp. 9393–9400, 2013.

34. L. He, H. S. Freeman, L. Lu, and S. Zhang, "Spectroscopic study of anthraquinone dye/amphiphile systems in binary aqueous/

organic solvent mixtures," Dyes & Pigments, vol. 91, no. 3, pp. 389–395, 2011·

35. I. Yumoto, K. Hirota, Y. Nodasaka, Y. Yokota, T. Hoshino, and K. Nakajima, "Alkalibacterium psychrotolerans sp. nov., a psychrotolerant obligate alkaliphile that reduces an indigo dye," International Journal of Systematic and Evolutionary Microbiology, vol. 54, no. 6, pp. 2379–2383, 2004. ·

36. C. P. Raut, M. D. Daley, K. K. Hunt et al., "Anaphylactoid reactions to isosulfan blue dye during breast cancer lymphatic mapping in patients given preoperative prophylaxis,"Journal of Clinical Oncology, vol. 22, no. 3, pp. 567–568, 2004.

37. Z.-F. Huang, J.-J. Zou, L. Pan, S. Wang, X. Zhang, and L. Wang, "Synergetic promotion on photoactivity and stability of $W_{18}O_{49}$/TiO_2 hybrid," Applied Catalysis B, vol. 147, pp. 167–174, 2014.

38. S. Sarkar and K. K. Chattopadhyay, "Visible light photocatalysis and electron emission from porous hollow spherical $BiVO_4$ nanostructures synthesized by a novel route,"Physica E, vol. 58, pp. 52–58, 2014.

39. V. L. Blair, E. J. Nichols, J. Liu, and S. T. Misture, "Surface modification of nanosheet oxide photocatalysts," Applied Surface Science, vol. 268, pp. 410–415, 2013.

40. Z. Li, Y. Shen, C. Yang et al., "Significant enchancement in the visible light pjotocatalytic properties of $BiFeO_3$-graphene nanohybrids," Journal of Materials Chemistry A, vol. 1, pp. 823–829, 2013.

41. C. J. Miller, H. Yu, and T. Waite, "Degradation of rhodamine B during visible light photocatalysis employing Ag@AgCl embedded on reduced graphene oxide," Colloids & Surface A, vol. 435, pp. 147–153, 2013.

42. S. Rashidi, M. Nikazar, A. V. Yazdi, and R. Fazaeli, "Optimized photocatalytic degradation of reactive blue 2 by TiO_2/UV process," Journal of Environmental Science and Health, Part A, vol. 49, pp. 452–462, 2014.

43. R.-H. Jie, G.-B. Guo, W.-G. Zhao, and S.-L. An, "Preparation and photocatalytic degradation of methyl orange of nano-powder TiO_2 by hydrothermal method supported on activated carbon," Journal of Synthetic Crystals, vol. 42, pp. 2144–2149, 2013.

44. Z. Bouberka, K. A. Benobbou, A. Khenifi, and U. Maschke, "Degradation by irradiation of an acid orange 7 on colloidal TiO_2/(LDHs)," Journal of Photochemistry and Photobiology A, vol. 275, pp. 21–29, 2014.

45. B. Pant, H. R. Pant, N. A. M. Barakat et al., "Incoporation of cadmium sulfide nanoparticles on the cadmium titanate nanofibers for enhanced organic dyes degradation and hydrogen release," Ceramics International, vol. 40, pp. 1553–1559, 2014.

46. H. Hagiwara, M. Nagatomo, C. Seto, S. Ida, and T. Ishihara, "Dye-modification effects on water splitting activity of GaN:ZnO photocatalyst," Journal of Photochemistry and Photobiology A, vol. 272, pp. 41–48, 2013.

47. A. Prasannan and T. Imae, "One-pot synthesis of fluorescent carbon dots from orange waste peels," Industrial & Engineering Chemistry Research, vol. 52, pp. 15673–15678, 2013.

48. L. G. Devi and M. L. Arunakumari, "Enhanced photocatalytic performance of Hemin (chloro(protoporhyinato) iron (III)) anchored TiO_2 photocatalyst for methyl orange degradation: a surface modification method," Applied Surface Science, vol. 276, pp. 521–528, 2013.

49. B. P. Nenavathu, A. V. R. Krishna Rao, A. Goyal, A. Kapoor, and R. K. Dutta, "Synthesis, characterization and enchanced photocatalytic degradation efficiency of Se doped ZnO nanoparticles using trypan blue as a model dye," Applied Catalysis A, vol. 459, pp. 106–113, 2013.

50. Y. Huo, Z. Xie, X. Wang, H. Li, M. Hoang, and R. A. Caruso, "Methyl orange removal by combined visible-light photocatalysis and membrane distillation," Dyes & Pigments, vol. 98, pp. 106–112, 2013.

51. H. U. Lee, G. Lee, J. C. Park et al., "Efficient visible-light responsive TiO_2 nanoparticles incoporated magnetic carbon photocatalysts," Chemical Engineering Journal, vol. 240, pp. 91–98, 2014.

52. S. O. Saheed, S. J. Modise, and A. M. Sipamla, "TiO_2 supported clinoptilotile: characterization and optimization of operational parameters for methyl orange removal," Advanced Materials Research, vol. 781, pp. 2249–2252, 2013.

53. S. J. Hu, J. Yang, and X. H. Liao, "Highly efficient degradation of methylene blue on microwave synthesized $FeVO_4$ nanoparticles

photocatalyst under visible-light irradiation," Applied Mechanics and Materials, vol. 372, pp. 153–157, 2013.

54. E. S. Aazam and R. M. Mohamed, "Environmental remediation of direct blue dyes solutions by photocatalytic oxidation with cuppor oxide," Journal of Alloys and Compounds, vol. 577, pp. 550–555, 2013.

55. M. Xu, J. Guo, Y. Cen, X. Zhong, W. Cao, and G. Sun, "Shewanella decolorationis sp. nov., a dye-decolorizing bacterium isolated from activated sludge of a waste-water treatment plant," International Journal of Systematic and Evolutionary Microbiology, vol. 55, no. 1, pp. 363–368, 2005·

56. E. Abadulla, T. Tzanov, S. Costa, K.-H. Robra, A. Cavaco-Paulo, and G. M. Gubitz, "Decolorization and detoxification of textile dyes with a laccase from Trametes hirsuta," Applied and Environmental Microbiology, vol. 66, no. 8, pp. 3357–3362, 2000

57. N. Sakkayawong, P. Thiravetyan, and W. Nakbanpote, "Adsorption mechanism of synthetic reactive dye wastewater by chitosan," Journal of Colloid and Interface Science, vol. 286, no. 1, pp. 36–42, 2005.

58. Y. Wang, G. Wang, H. Wang, C. Liang, W. Cai, and L. Zhang, "Chemical-template synthesis of micro/nanoscale magnesium silicate hollow spheres for waste-water treatment," Chemistry: A European Journal, vol. 16, no. 11, pp. 3497–3503, 2010.

59. J.-L. Gong, B. Wang, G.-M. Zeng et al., "Removal of cationic dyes from aqueous solution using magnetic multi-wall carbon nanotube nanocomposite as adsorbent," Journal of Hazardous Materials, vol. 164, no. 2-3, pp. 1517–1522, 2009.

60. J. M. Peralta-Hernández, Y. Meas-Vong, F. J. Rodríguez, T. W. Chapman, M. I. Maldonado, and L. A. Godínez, "Comparison of hydrogen peroxide-based processes for treating dye-containing wastewater: decolorization and destruction of Orange II azo dye in dilute solution," Dyes & Pigments, vol. 76, no. 3, pp. 656–662, 2008·

61. B. N. Joshi, H. Yoon, S.-H. Na, J.-Y. Choi, and S. S. Yoon, "Enchanced photocatalytic performance of graphene-ZnO nanoplatet composite thin films prepared by electrostatic spray deposition," Ceramics International, vol. 40, pp. 3647–3654,

2014.

62. M. Wang, J. Huang, Z. Tong, W. Li, and J. Chen, "Reduced graphene oxide-cuprous oxide composite via facial deposition for photocatalytic dye-degradation," Journal of Alloys and Compounds, vol. 568, pp. 26–35, 2013.

63. U. G. Akpan and B. H. Hameed, "Development and photocatalytic activities of TiO_2 doped with Ca-Ce-W in the degradation of acid red 1 under visible light irradiation,"Desalination & Water Treatment, 2013.

64. S. Zhang, J. Li, M. Zeng et al., "In situ synthesis of water-soluble magnetic graphitic carbon nitride photocatalyst and its synergistic catalytic performance," ACS Applied Materials & Interfaces, vol. 5, pp. 1235–1243, 2013.

65. P. Peralta-Zamora, "Photoelectrochemical or electrophotochemical processes?" Journal of the Brazilian Chemical Society, vol. 21, no. 9, pp. 1621–1625, 2010.

66. M. E. Olya and A. Pirkarami, "Cost-effective photoelectrocatalytic treatment of dyes in a batch reactor equipped with solar cells," Separation & Purification Technology, vol. 118, pp. 557–566, 2013.

67. J. Ma, B. Cui, J. Dai, and D. Li, "Mechanism of adsorption of anionic dye from aqueous solutions onto organobentonite," Journal of Hazardous Materials, vol. 186, no. 2-3, pp. 1758–1765, 2011.

68. M.-X. Zhu, L. Lee, H.-H. Wang, and Z. Wang, "Removal of an anionic dye by adsorption/precipitation processes using alkaline white mud," Journal of Hazardous Materials, vol. 149, no. 3, pp. 735–741, 2007.

69. Y. Dong, W. Dong, C. Liu, Y. Chen, and J. Hua, "Photocatalytic decoloration of water-soluble azo dyes by reduction based on bisulfite-mediated borohydride," Catalysis Today, vol. 126, no. 3-4, pp. 456–462, 2007

70. A. Szyguła, E. Guibal, M. A. Palacín, M. Ruiz, and A. M. Sastre, "Removal of an anionic dye (Acid Blue 92) by coagulation-flocculation using chitosan," Journal of Environmental Management, vol. 90, no. 10, pp. 2979–2986, 2009

71. M. A. M. Salleh, D. K. Mahmoud, W. A. W. A. Karim, and A. Idris, "Cationic and anionic dye adsorption by agricultural solid wastes: a comprehensive review,"Desalination, vol. 280, no. 1–3, pp. 1–13, 2011·

72. G. A. Ikhtiyarova, A. S. Özcan, O. Gök, and A. Özcan, "Characterization of natural- and organo-bentonite by XRD, SEM, FT-IR and thermal analysis techniques and its adsorption behaviour in aqueous solutions," Clay Minerals, vol. 47, pp. 31–44, 2012.

73. P. Baskaralingam, M. Pulikesi, D. Elango, V. Ramamurthi, and S. Sivanesan, "Adsorption of acid dye onto organobentonite," Journal of Hazardous Materials, vol. 128, no. 2-3, pp. 138–144, 2006. ·

74. J. F. Ma, J. M. Yu, B. Y. Cui, D. L. Li, and J. Dai, "Treatment of dye wastewater by zero valent iron composited organobentonite," Advanced Materials Research, vol. 340, pp. 229–235, 2011

75. I. L. Lagadic, M. K. Mitchell, and B. D. Payne, "Highly effective adsorption of heavy metal ions by a thiol-functionalized magnesium phyllosilicate clay," Environmental Science and Technology, vol. 35, no. 5, pp. 984–990, 2001. ·

76. F. Gao, P. Botella, A. Corma, J. Blesa, and L. Dong, "Monodispersed mesoporous silica nanoparticles with very large pores for enhanced adsorption and release of DNA,"Journal of Physical Chemistry B, vol. 113, no. 6, pp. 1796–1804, 2009. ·

77. X. Wang, R. Liu, M. M. Waje et al., "Sulfonated ordered mesoporous carbon as a stable and highly active protonic acid catalyst," Chemistry of Materials, vol. 19, no. 10, pp. 2395–2397, 2007·

78. T. Suteewong, H. Sai, R. Cohen et al., "Highly aminated mesoporous silica nanoparticles with cubic pore structure," Journal of the American Chemical Society, vol. 133, no. 2, pp. 172–175, 2011·

79. J. Tao, W. Jiang, H. Zhai, H. Pan, R. Xu, and R. Tang, "Structural components and anisotropic dissolution behaviors in one hexagonal single crystal of -tricalcium phosphate," Crystal Growth and Design, vol. 8, no. 7, pp. 2227–2234, 2008. · ·

80. H. Pan, J. Tao, R. Xu, and R. Tang, "Adsorption processes of gly and glu amino acids on hydroxyapatite surfaces at the atomic level," Langmuir, vol. 23, no. 17, pp. 8972–8981, 2007. ·

81. S. Rashmi and V. Preeti, "Decolorisation of aqueous dye solutions by low-cost adsorbents: a review," Coloration Technology, vol. 129, pp. 85–108, 2013.

82. F. H. Hussein, "Effect of photocatalytic treatments on physical and biological properties of textile dyeing wastewater," Asian Journal of Chemistry, vol. 25, pp. 9387–9392, 2013.

83. J. Moon, C. Y. Yun, K.-W. Chung, M.-S. Kang, and J. Yi, "Photocatalytic activation of TiO_2 under visible light using Acid Red 44," Catalysis Today, vol. 87, no. 1–4, pp. 77–86, 2003·

84. M. Y. Abdelaal and R. M. Mohamed, "Novel Pd/TiO_2 nanocomposite prepared by modified sol-gel method for photocatalytic degradation of methylene blue dye under visible light irradation," Journal of Alloys and Compounds, vol. 576, pp. 201–207, 2013.

85. M. G. Weinbauer, C. Beckmann, and M. G. Höfle, "Utility of green fluorescent nucleic acid dyes and aluminum oxide membrane filters for rapid epifluorescence enumeration of soil and sediment bacteria," Applied and Environmental Microbiology, vol. 64, no. 12, pp. 5000–5003, 1998.

86. A. Hamrouni, H. Lachheb, and A. Houas, "Synthesis, characterization and photocatalytic activity of ZnO-SnO_2 nanocomposites," Materials Science and Engineering B, vol. 178, pp. 1371–1379, 2013.

87. H. G. Cha, H. S. Noh, M. J. Kang, and Y. S. Kang, "Photocatalysis: progress using manganese-doped hematite nanocrystals," New Journal of Chemistry, vol. 37, pp. 4004–4009, 2013.

88. T. Jiang, L. Zhang, M. Ji et al., "Carbon nanotubes/TiO_2 nanotubes composite photocatalysts for efficient degrdation of methyl orange dye," Particuology, vol. 11, pp. 737–742, 2013.

89. A. Fatemeh, F. Nazanin, and M. A. Tehrani Ramin, "Preparation of NiO loaded on TiO_2 nanostructure as nanophotocatalyst and its photocatalytic activity for degradation of methylene blue," Research Journal of Chemistry and Environment, vol. 17, pp. 92–96, 2013.

90. F. A. Harraz, R. M. Mohamed, M. M. Rashad, Y. C. Wang, and W. Sigmund, "Magnetic nanocomposite based on titania-silica/cobalt ferrite for photocatalytic degradation of methylene blue dye," Ceramics International, vol. 40, pp. 375–384, 2014.

91. J. Zhang, W. Liu, X. Wang, X. Wang, B. Hu, and H. Liu, "Enchanced decolarization activity by $Cu_2O@TiO_2$ nanobelts heterostructures via a strong adsorption-weak photodegradation process," Applied Surface Science, vol. 282, pp. 84–91, 2013.

92. M. Hamadanian, M. Behpour, A. S. Razavian, and V. Habbari, "Structural, morphological and photocatalytic characterisations of Ag-coated anatase TiO_2 fabricated by the sol gel dip coating method," Journal of Experimental Nanoscience, vol. 8, pp. 901–912, 2013.

93. B. Shahmoradi, A. Maleki, and K. Byrappa, "Removal of dispersed orange 25 using in situ surface iron-doped TiO_2 nanoparticles," Desalination & Water Treatment, 2013.

94. Priyanka and V. C. Srivastava, "Photocatalytic oxidation of dye bearing wastewater by iron doped zinc-oxide," Industrial & Engineering Chemistry Research, vol. 52, no. 50, pp. 17790–17799, 2013. ·

95. L. Sun, Y. Shi, B. Li, X. Li, and Y. Wang, "Preparation and characterization of polypyyrole/TiO_2 nanocomposites by reverse microemulsion polymerization and its photocatalytic activity for the degradation of methyl orange under natural light,"Polymer Composites, vol. 34, no. 7, pp. 1076–1080, 2013

96. B. Saygi and D. Tekin, "Photocatalytic degradation kinetics of reactive black 5 (RB 5) dyestuff on TiO_2 modified by pretreatment with untrasound energy," Reaction Kinetics, Mechanisms and Catalysis, vol. 110, no. 1, pp. 251–258, 2013. ·

97. C. C. Pei and W. W. F. Leung, "Photocatalytic degradation of Rhodamine B by TiO_2/ZnO nanofibers under visible-light irradiation," Separation & Purification Technology, vol. 114, pp. 108–116, 2013.

98. A. F. Shojaei, A. R. Tabari, and M. H. Loghmani, "Normal spinel $CoCr_2O_4$ and $CoCr_2O_4$/TiO_2 nanocomposite as novel photocatalysts for degradation of dyes," Micro & Nano Letters, vol. 8, pp. 426–431, 2013.

99. J. B. Joo, I. Lee, M. Dahl, G. D. Moon, F. Zaera, and Y. Yin, "Controllable synthesis of mesoporous TiO$_2$ hallows shells: toward an efficient photocatalyst," Advanced Functional Materials, vol. 23, pp. 4246–4254, 2013.

100. L. Shi, L. Liang, J. Ma et al., "Highly efficient visible light-driven Ag/AgBr/ZnO composite photocatalyst for degrading Rhodamine B," Ceramics International, vol. 40, pp. 3495–3502, 2014. ·

101. B. M. Rajbongshi and S. K. samdarshi, "ZnO and Co-ZnO nanorods-Complementary role of oxygen vacancy in photocatalytic activity of under UV and visible radiation flux,"Materials Science and Engineering B, vol. 182, pp. 21–28, 2014.

102. P. Muthirulan, C. K. Nirmala Devi, and M. M. Sundaram, "Facile synthesis of novel hierarchiral TiO$_2$@Poly(o-phenylenediamine) core-shell structures with enchanced photocatalytic performance under solar light," Journal of Environmental Chemical Engineering, vol. 1, pp. 620–627, 2013.

103. W. Zhang, T. Hu, B. Yang, P. Sun, and H. He, "The effect of boron content on properties of B-TiO$_2$ photocatalyst prepared by sol-gel method," Journal of Advanced Oxidation Technologies, vol. 16, pp. 261–267, 2013.

104. A. Mohamed, S. Alberto, M.-F. Victor, and E. Luis, "Removal of basic yellow cationic dye by an aqueous dispersion of Moroccan stevensite," Applied Clay Science, vol. 80, pp. 46–51, 2013.

105. Z. D. Meng, L. Zhu, S. Ye et al., "Heterogenous photocatalytic degradation of anionic and cationic dyes over fe-fullerence/TiO$_2$ under visible light," Asian Journal of Chemistry, vol. 25, pp. 6001–6007, 2013.

106. V. V. Panic and S. J. Velickovic, "Removal of model cationic dye by adsorption onto poly(methacrylic acid)/zeolite hydrogel composites: kinetics, equilibrium study and image analysis," Separation & Purification Technology, vol. 122, pp. 284–294, 2014. ·

107. Q. Li, Q.-Y. Yue, H.-J. Sun, Y. Su, and B.-Y. Gao, "A comparative study on the properties, mechanisms and process designs for the adsorption of non-ionic or anionic dyes onto cationic-polymer/bentonite," Journal of Environmental Management, vol. 91, no. 7, pp. 1601–1611, 2010.

108. Y. Zhen, Y. Hu, J. Ziwen et al., "Flocculation of both anionic and cationic dyes in aqueous solutions by the amphoteric grafting flocculant carboxymethyl chitosan-graft-polyacrylamide," Journal of Hazardous Materials, vol. 254, pp. 36–45, 2013.

109. A. Afkhami, M. Saber-Tehrani, and H. Bagheri, "Modified maghemite nanoparticles as an efficient adsorbent for removing some cationic dyes from aqueous solution,"Desalination, vol. 263, no. 1–3, pp. 240–248, 2010. ·

110. S. Iyyapushpam, S. T. Nishanti, and D. Pathinettam Padiyan, "Photocatalytic degradation of methyl orange using -Bi_2O_3 prepared without surfactant," Journal of Alloys and Compounds, vol. 563, pp. 104–107, 2013. ·

111. S. K. Kansal, R. Lamba, S. K. Mehta, and A. Umar, "Photocatalytic degradation of Alizarin Red S using simply synthesized ZnO nanoparticles," Materials Letters, vol. 106, pp. 385–389, 2013.

112. C. Zhu, Y. Li, Q. Su et al., "Electrospinning direct preparation of SnO_2/Fe_2O_3heterojunction nanotubes as an efficient visible-light photocatalyst," Journal of Alloys and Compounds, vol. 575, pp. 333–338, 2013.

113. S. Balachandran, S. G. Praveen, R. Velmurugan, and M. Swaminathan, "Facile fabrication of highly efficient, reusable heterostructured Ag-ZnO-CdO and its twin applications of dye degradation under natural sunlight and self-cleaning," RSC Advances, vol. 4, pp. 4353–4362, 2014.

114. H. Gulce, V. Eskizeybek, B. Haspulat, F. Sari, A. Gulce, and A. Avci, "Preparation of a new polyaniline/CdO nanocomposite and investigation of its photocatalytic activity: comparative study under UV light and natural sunlight irradation," Industrial & Engineering Chemistry Research, vol. 52, pp. 10924–10934, 2013.

115. E. Repo, S. Rengaraj, S. Pulkka et al., "Photocatalytic degradation of dyes by CdS micropoheres under near UV and blue LED radiatin," Separation & Purification Technology, vol. 120, pp. 206–214, 2013.

116. N. Divya, A. Bansal, and A. K. Jana, "Nano-photocatalysts in the treatment of colored wastewater—a review," Materials Science Forum, vol. 734, pp. 349–363, 2013.

117. D. Pathania, S. Sarita, and B. S. Rathore, "Synthesis, characterization and photocatalytic application of bovine serum albumin capped cadmium sulphide nanopartilces,"Chalcogenide Letters, vol. 8, no. 6, pp. 396–404, 2011.

118. A. Tadjorodi, M. Imani, and H. Kerdari, "Experimental design to optimize the synthesis of CdO cauliflower-like nanostructure and high performance in photodegaradtion of toxic azo dyes," Materials Research Bulletin, vol. 48, pp. 935–942, 2013.

119. L. Bouna, B. Rhouta, and F. Maury, "Physicochemical study of photocatalytic activity of TiO_2supported polygorskite clay mineral," International Journal of Photoenergy, vol. 2013, Article ID 815473, 6 pages, 2013. ·

120. Z.-L. Ma, G.-F. Huang, D.-S. Xu, M.-G. Xia, W.-Q. Huang, and Y. Tian, "Coupling effect of la doping and porphyrin sensitization on photocatalytic activity of nanocrystalline TiO_2," Materials Letters, vol. 108, pp. 37–40, 2013.

121. Z. Shi, M. Zhou, D. Zheng, H. Liu, and S. Yao, "Preparation of Ce-doped TiO_2 hollow fibers and their photocatalytic degradation properties for dye compound," Journal of the Chinese Chemical Society, vol. 60, pp. 1156–1162, 2013.

122. N. B. Gusiak, I. M. Kobasa, and S. S. Kurek, "Nature inspired dyes for the sensitization of titanium dioxide photocatalyst," Chemik, vol. 67, pp. 1191–1198, 2013.

123. Y. Huo, X. Chen, J. Zhang, G. Pan, J. Jia, and H. Li, "Ordered macroporous Bi_2O_3/TiO_2film coated on a rotating disk with enchanced photocatalytic activity under visible irradiation," Applied Catalysis B, vol. 148, pp. 550–556, 2014.

124. L.-J. Kim, J.-W. Jang, and J.-W. Park, "Nano TiO_2 functionalized magnetic-cored dendrimer as a photocatalyst," Applied Catalysis B, vol. 147, pp. 973–979, 2014.

125. J. Rattanarak, W. Mekprasart, W. Pecharapa, and W. Techitdheera, "Photocatalytic activities under UV light of ball-milled TiO_2 photocatalyts," Advanced Materials Research, vol. 802, pp. 237–241, 2013.

126. W. C. Liu, H. Y. Xu, T. N. Shi, L. C. Wu, and P. Li, "Preparation and photocatalytic activity of TiO_2/tourmaline composite catalyst," Advanced Materials Research, vol. 800, pp. 464–470, 2013.

127. T. E. Agustina, F. S. Arsyad, and M. Abdullah, "Photocatalytic degradation of C.I reactive red 2 by using TiO_2-coated PET plastic under solar irradiation," Advanced Materials Research, vol. 789, pp. 180–188, 2013.

128. X. Sun, C. Li, L. Ruan et al., "Ce-doped SiO_2@TiO_2 nanocomposite as an effective visible light photocatalyst," Journal of Alloys and Compounds, vol. 585, pp. 800–804, 2014.

129. M.-C. Wu, H.-C. Liao, Y.-C. Cho et al., "Photocatalytic activity of nitrogen-doped TiO_2-based nanowires: a photo-assisted Kelvin probe force microcopy study," Journal of Nanoparticle Research, vol. 16, pp. 1–11, 2014.

130. S. Shi, M. A. Gondal, A. A. Al-Saadi et al., "Facile preparation of g-C_3N_4 modified BiOCl hybrid photocatalyst and vital role of frontier orbital energy levels of model compounds in photoactivity enchecement," Journal of Colloid and Interface Science, vol. 416, pp. 212–219, 2014.

131. A. E. Nogueira, E. Longo, E. R. Leite, and E. R. Camargo, "Synthesis and photocatalytic properties of bismuth titanate with different structures via oxidant peroxo method (OPM)," Journal of Colloid and Interface Science, vol. 415, pp. 89–94, 2014.

132. Q. Wang, J. Hui, L. Yang et al., "Enchanced photoactivity performance of Bi_2O_3/H-ZSM-5 composite for rgodamine B degradation under UV light irradiation," Applied Surface Science, vol. 289, pp. 224–229, 2014. ·

133. Y. L. Ma, R. S. Xue, and H. S. Yan, "Photo-degaradtion alkali lignin by Bis-(2-Methyl Quinoline) squarylium cyanine TiO_2 photocatalyst in sunlight," Advanced Materials Research, vol. 838, pp. 2717–2720, 2014.

134. Q. Wang, J. Hui, Y. Huang et al., "The preparation of BiOCl photocatalyst and its performance of photodegradation on dyes," Materials Science in Semiconductor Processing, vol. 17, pp. 87–93, 2014.

135. W. T. Yi and C. Y. Yan, "A novel visible-light-driven photocatalyst: Pt surface modofied Bi_2WO_6-WO_3 composite," Applied Mechanics and Materials, vol. 448, pp. 178–181, 2014.

136. Z. Liu, B. Wu, J. Niu, X. Huang, and Y. Zhu, "Solvothermal synthesis of BiOBr thin film and its photocatalytic performance," Applied Surface Science, vol. 288, pp. 369–372, 2014.

137. Z. Li, S. Yang, J. Zhou et al., "Novel mesoporous g-C_3N_4 and $BiPO_4$ nanorods hybrid architectures and their enchanced visible-light-driven photocatalytic performances,"Chemical Engineering Journal, vol. 241, pp. 344–351, 2014. ·

138. M. Wang, Y. Che, C. Niu, M. Dang, and D. Dong, "Effective visible light-active boron and europium co-doped $BiVO_4$ synthesized by sol-gel method for photodegradation of methyl orange," Journal of Hazardous Materials, vol. 262, pp. 447–455, 2013. ·

139. G. Zhu, M. Hojamberdiev, K. I. Katsumata et al., "Heterostructured $Fe_3O_4/Bi_2O_2CO_3$photocatalyst: synthesis, characterization and application in recycleable photodegradation of organic dyes under visible light irradiation," Materials Chemistry and Physics, vol. 142, pp. 95–105, 2013.

140. Q. Wang, J. Hui, J. Li et al., "Photodegradation of methyl orange with PANI-modified BiOCl photocatalyst under visble light irradiation," Applied Surface Science, vol. 283, pp. 577–583, 2013.

141. Z. Zhanying, I. M. O'Hara, A. K. Geoff, and O. S. D. William, "Comparative study on adsorption of two cationic dyes by milled sugarcane bagasse," Industrial Crops and Products, vol. 42, pp. 41–49, 2013

142. S.-K. Mousa, A. Mokhtar, and G. Kamaladin, "Preparation of chitosan-ethyl acrylate as a biopolymer adsorbent for basic dyes removal from colored solutions," Journal of Environmental Chemical Engineering, vol. 1, pp. 406–415, 2013.

143. K. Turhan, I. Durukan, S. A. Ozturkcan, and Z. Turgut, "Decolorization of textile basic dye in aqueous solution by ozone," Dyes & Pigments, vol. 92, no. 3, pp. 897–901, 2012.

144. A. T. Kah, M. Norhashimah, T. T. Tjoon, N. Ismail, and P. Panneerselvam, "Removal of cationic dye by magnetic nanoparticle (Fe_3O_4) impregnated onto activated maize cob powder and kinetic study of dye waste adsorption," APCBEE Procedia, vol. 1, pp. 83–89, 2012.

145. S. M. R. Billah, R. M. Christie, and R. Shamey, "Direct coloration of textiles with photochromic dyes. Part 1: application of spiroindolinonaphthoxazines as disperse dyes to polyester, nylon and acrylic fabrics," Coloration Technology, vol. 124, no. 4, pp. 223–228, 2008

146. M. M. Hassan and C. J. Hawkyard, "Decolourisation of aqueous dyes by sequential oxidation treatment with ozone and Fenton's reagent," Journal of Chemical Technology and Biotechnology, vol. 77, no. 7, pp. 834–841, 2002. ·

147. B. Mralidharan and S. Laya, "A new approach to dyeing of 80:20 polyester/cotton blended fabric using disperse and reactive dyes," ISRN Materials Science, vol. 2011, Article ID 907493, 12 pages, 2011. ·

148. H. Y. Ze, F. T. Jing, Q. W. Xiao, C. Xiu, L. Wei, and Z. Ying, "Effects of disperse dyes on dyeing of ethylated Chinese fir powder," Advanced Materials Research, vol. 788, pp. 241–245, 2013.

149. W. Cui, H. Wang, L. Liu, Y. Liang, and J. G. McEvoy, "Plasmonic Ag@AgCl-intercalated $K_4Nb_6O_{17}$ composite for enchanced photocatalytic degradation of Rhodamine B under visible light," Applied Surface Science, vol. 283, pp. 820–827, 2013.

150. Z. Ali, N. R. Khalid, M. Nawaz Chaudhry, S. Tajammul Hussain, I. Ahamad, and N. A. Niaz, "Significant effect of graphene on catalytic degradation of methylene blue by pure and Ce doped TiO_2 at nanoscale," Digest Journal of Nanomaterials and Biostructures, vol. 8, pp. 1525–1534, 2013.

151. Q. Wang, J. Li, Y. Bai et al., "Photodegradation of textile dye Rhodamine B over a novel biopolymer-metal complex wool-Pd/CdS photocatalysts under visible light irradiation,"Journal of Photochemistry and Photobiology B, vol. 126, pp. 47–50, 2013.

152. B. Krishnakumar and M. Swaminathan, "Solar photocatalytic degradation of Naphthol Blue Black," Desalination & Water Treatment, vol. 51, pp. 6572–6579, 2013.

153. O. Sharma and M. K. Sharma, "Use of cobalt hexacyanoferrate(II) semiconductor in photocatalytic degradation of neutral red dye," International Journal of ChemTech Research, vol. 5, pp. 1615–1622, 2013.

154. P. Du, L. Song, J. Xiong, L. Wang, and N. Li, "A photovoltaic smart textile and a photocatalytic functional textile based on co-electronspun TiO_2/MgO core-sheath nanorods: novel textile of integrating energy and environmental science with textile research," Textile Research Journal, vol. 83, pp. 1690–1702, 2013.

155. J. Yang, C. Chen, H. Ji, W. Ma, and J. Zhao, "Mechanism of TiO_2-assisted photocatalytic degradation of dyes under visible irradiation: photoelectrocatalytic study by TiO_2-film electrodes," Journal of Physical Chemistry B, vol. 109, no. 46, pp. 21900–21907, 2005. ·

156. P. Eskandari, F. Kazemi, and Y. Azizian-Kalandaragh, "Convenient preparation of CdS nanostructures as a highly efficient photocatalyst under blue LED and solar light irradiation," Separation & Purification Technology, vol. 120, pp. 180–185, 2013.

157. F. Wen and C. Li, "Hybrid artificial photosynthetic systems comprising semiconductors as light harvesters and biomimetic complexes as molecular cacatalyst," Accounts of Chemical Research, vol. 46, pp. 2355–2364, 2013.

158. L. M. Duan, J. H. Liu, Q. Y. Pang, L. Xu, and Z. R. Liu, "Efficient sunlight active nanocomposite photocatalytst for degradation of pollutant organic dyes," Advanced Materials Research, vol. 726, pp. 650–653, 2013.

159. S. Da Dalt, A. K. Alves, and C. P. Bergmann, "Photocatalytic degradation of methyl orange in water solutions in the presence of $MWCNT/TiO_2$ composites," Materials Research Bulletin, vol. 48, pp. 1845–1850, 2013.

160. N. Divya, A. Bansal, and A. K. Jana, "Photocatalytic degradation of azo Orange II in aqueous solutions using copper-impregnated titania," International Journal of Environmental Science and Technology, vol. 10, pp. 1265–1274, 2013.

161. J. Dong, H. Xu, F. Zhang, C. Chen, L. Liu, and G. Wu, "Synergistic effect over photocatalytic active Cu_2O thin films and their morphological and orientational transformation under visible light irradaiation," Applied Catalysis A, vol. 470, pp. 294–302, 2014.

162. O. Sharma and M. K. Sharma, "Copper hexacyanoferrate (II) as photocatalyst: decolarisation of neutral red dye," International Journal of ChemTech Research, vol. 5, pp. 2706–2716, 2013.

163. S. S. Shinde, C. H. Bhosale, and K. Y. Rajpure, "Kinetic analysis of heterogenous photocatalysis: role of hydroxyl radicals," Catalysis Review, vol. 55, pp. 79–133, 2013.

164. K. Zhou, Y. Shi, S. Jiang, Y. Hu, and Z. Gui, "Facile preparation of Cu_2O/carbon heterostucture with high photocatalytic activity," Materials Letters, pp. 213–216, 2013.

165. X. Lu, N. Hu, J. Li, H. Ma, K. Du, and R. Zhao, "Influence of TiO impregnated with a novel copper (II) carboxylic porphyrin and its application in photocatalytic degradation of 4-nitrophenol," Research on Chemical Intermediates, vol. 40, no. 5, pp. 1911–1922, 2014. ·

166. W. Hu, F. Ren, R. Bai, Z. Zhou, and W. Xu, "Preparation and photocatalytic properties of CuO-TiO_2/conductive polymer fiber composites," Acta Scientiae Circumstantiae, vol. 33, pp. 431–436, 2013.

167. P. Deepak and S. Shikha, "Effect of surfactants and electrolyte on removal and recovery of basic dye by using ficus carica cellulosic fibers as biosorbent," Surfactant, vol. 49, pp. 306–314, 2011.

168. R. O. Cristavoa, A. P. M. Tavares, J. M. Loureiro, R. A. R. Boaventura, and E. A. Macedo, "Optimisation of reactive dye degradation by laccase using Box-Behnken design,"Environmental Technology, vol. 29, no. 12, pp. 1357–1364, 2008. ·

169. A. Khan, N. A. Mir, M. M. Haque, M. Muneer, and C. Boxall, "Solar photocatalytic decolorization of two azo dye derivatives, acid red 17 and reactive red 241 in aqueous suspension," Science of Advanced Materials, vol. 5, pp. 160–165, 2013.

170. S. K. Kavitha and P. N. Palanisamy, "Solar photocatalitic degradation of Vat Yellow 4 dye in aqueous suspension of TiO_2-optimization of operational parameters," Advances in Environmental Sciences, vol. 2, pp. 189–202, 2010.

171. S. T. Tan, A. A. Umar, A. Balouch et al., "ZnO nanocubes with (1 0 1) basal plane photocatalyst prepared via a low-frequency ultrasonic assisted hydrolysis process,"Untrasonic Sonochem, vol. 21, pp. 754–760, 2014.

172. A. A. Taha, A. A. Hriez, Y.-N. Wu, H. Wang, and F. Li, "Direct synthesis of novel vanadium oxide embedded porous carbon nanofiber decorated with iron nanoparticles as a low-cost and highly efficient visible-light-driven photocatalyst," Journal of Colloid and Interface Science, vol. 417, pp. 199–205, 2014.

173. H. Xu, J. Zhang, Y. Chen, H. Lu, J. Zhuang, and J. Li, "Synthesis of polyaniline-modified MnO_2 composite nanorods and their

photocatalytic application," Materials Letters, vol. 117, pp. 21–23, 2014.

174. C. Wang, X. Zhang, B. Yuan et al., "Multi-heterojunction photocatalyst based on WO_3 nanorods: structural design and optimization for enchanced photocatalytic activity under visible light," Chemical Engineering Journal, vol. 237, pp. 29–37, 2014.

175. N. A. S. Al-Areqi, A. Al-Alas, A. S. N. Al-Kamali, K. A. S. Ghaleb, and K. Al-Mureish, "Photodegradation of 4-SPPN dye catalyzed by Ni(II)-substited $Bi_2VO_{5.5}$ system under visible light irradiation: influence of phase stability and perovskite vanadate oxygen vacancies of photocatalyst," Journal of Molecular Catalysis A, vol. 381, pp. 1–8, 2014. ·

176. Y. He, D. Li, J. Chen et al., "Sn_3O_4: a novel heterovalent-tin photocatalyst with hierarchical 3D nanostructures under visible light," RSC Advances, vol. 4, pp. 1266–1269, 2014.

177. O. F. Lopes, E. C. Paris, and C. Ribeiro, "Synthesis of Nb_2O_5 nanoparticles through the oxidant peroxide method applied to organic pollutant photodegradation: a mechanistic study," Applied Catalysis B, vol. 144, pp. 800–808, 2014.

178. M. Buchalska, J. Kuncewicz, E. Swietek et al., "Photoinduced hole injection in semiconductor-coordination compoun system," Coordination Chemistry Reviews, vol. 257, pp. 767–775, 2013.

179. A. Abidov, B. Allebergenov, O. Tursunkulov et al., "The evaluation of photocatalytic properties of iron doped titania photocatalyst by degradation of methylene blue using fluorescent light source," Advanced Materials Research, vol. 652, pp. 1700–1703, 2013.

180. M. Luo, D. Bowden, and P. Brimblecombe, "Removal of dyes from water using a TiO_2 photocatalyst supported on black sand," Water, Air, and Soil Pollution, vol. 198, no. 1–4, pp. 233–241, 2009·

181. H. M. Lim, J. S. Jung, D. S. Kim, D. J. Lee, S.-H. Lee, and W. N. Kim, "Modification of natural zeolite powder and its application to interior non-woven textile for indoor air quality control," Materials Science Forum, vol. 510-511, pp. 934–937, 2006.

182. R. Rahimi, M. M. Moghaddas, S. Zargari, and R. Rahimi, "Synthesis of mesoporous V-TiO_2 with different surfactants: the effect of surfactant type on photocatalytic process," Advanced Materials Research, vol. 702, pp. 56–61, 2013.

183. L. Pinho, J. C. Hernandez-Garrido, J. J. Calvino, and M. J. Mosquera, "2D and 3D characterization of a surfactant-synthesized TiO_2-SiO_2 mesoporous photoctalytst obtained at ambient temperature," Physical Chemistry Chemical Physics, vol. 15, pp. 2800–2808, 2013.

184. H. Park, Y. Park, W. Kim, and W. Choi, "Surface modification of TiO_2 photocatalyst for environmental applications," Journal of Photochemistry and Photobiology C, vol. 15, pp. 1–20, 2013.

185. P. Goswami, R. K. Debnath, and J. N. Ganguli, "Photophysical and photochemical properties of nanosized cobalt-doped TiO_2 photocatalyst," Asian Journal of Chemistry, vol. 25, pp. 7118–7124, 2013.

186. H. Meng, B. Wang, S. Liu, R. Jiang, and H. Long, "Hydrothermal preparation, characterization and photocatalytic activity of TiO_2/Fe-TiO_2 composite catalysts,"Ceramics International, vol. 39, pp. 5785–5793, 2013.

187. S. Vivekanandhan, M. Schreiber, C. Mason, A. K. Mohanty, and M. Misra, "Maple leaf (Acer sp.) extract mediated green process for the functionalization of ZnO powders with silver nanoparticles," Colloids and Surface B, vol. 113, pp. 169–175, 2014.

188. S. Xie, Y. Liu, Z. Chen, X. Chen, and X. Wang, "Superior photocatalytic properties of phosphorous-doped ZnO nanocombs," RSC Advances, vol. 3, pp. 26080–26085, 2013.

189. S. Balachandran, K. Selvam, B. Babu, and M. Swaminathan, "The simple hydrothermal synthesis of Ag-ZnO-SnO_2 nanochain and its multiple applications," Dalton Transactions, vol. 42, pp. 16365–16374, 2013.

190. J. Miao, Z. Jia, H.-B. Lu, D. Habibi, and L.-C. Zhang, "Heterogenous photocatalytic degradation of mordant black 11 with ZnO nanoparticles under UV-Vis light," Journal of the Taiwan Institute of Chemical Engineers, 2013. ·

191. Q.-L. Ma, R. Xiong, B.-G. Zhai, and Y. M. Huang, "Core-shelled Zn/ZnO microspheres synthesised by ultrasonic irradation for photocatalytic applications," Micro & Nano Letters, vol. 8, pp. 491–495, 2013.

192. S. Ameen, A. M. Shaheer, H.-K. Seo, and H.-S. Shin, "Mineralization of rhodamine 6G dye over rose flower-like ZnO nanomaterials," Materials Letters, vol. 113, pp. 20–24, 2013.

193. N.-F. Hsu, M. Chang, and K.-T. Hsu, "Rapid synthesis of ZnO dandelion-like nanostructures and their applications in humidity sensing and photocatalysis,"Materials Science in Semiconductor Processing, vol. 21, pp. 200–205, 2014. ·

194. N. W. C. Jusoh, A. A. Jalil, S. Triwahyono et al., "Sequential desilication-isomorphous substitution route to prepare mesostructured silica nanoparticles loaded with ZnO and their photocatalytic activity," Applied Catalysis A, vol. 468, pp. 276–287, 2013.

195. L. M. Duan, J. H. Liu, X. T. Xu, L. Xu, and Z. R. Liu, "The preparation and sunlight activity of nanocomposite photocatalysts for degradation of methyl orange solution,"Advanced Materials Research, vol. 750, pp. 1397–1400, 2013.

196. Y. V. Marathe and V. S. Shrivastava, "Effective removal of non-biodegradable methyl orange dye by using CdS/activated carbon nanocomposite as a photocatalyst,"Desalination & Water Treatment, 2013. ·

197. G. Yang, B. Yang, T. Xiao, and Z. Yan, "One-step solvothermal synthesis of hierarchically porous nanostructured CdS/TiO_2 heterojunction with higher visible light photocatalytic activity," Applied Surface Science, vol. 283, pp. 402–410, 2013.

198. M. Liu, J. Zheng, Q. Liu et al., "The preparation, load and photocatalytic performance of N-doped and CdS-coupled TiO_2," RSC Advabces, vol. 3, pp. 9483–9489, 2013.

199. J. Fu, B. Chang, Y. Tian, F. Xi, and X. Dong, "Novel C_3N_4-CdS composite photocatalysts with organic-inorganic heterojunctions: in situ synthesis, exceptional activity, high stability and photocatalytic mechanism," Journal of Materials Chemistry A, vol. 1, pp. 3083–3090, 2013.

200. L. Shao, G. Xing, W. Lv et al., "Photodegradation of azo-dyes in aqueous solution by polyacrylonitrile nanofiber mat-supported metalloporphyrins," Polymer International, vol. 62, pp. 289–294, 2013.

201. N. Sobana, B. Krishnakumar, and M. Swaminathan, "Synergism and effect of operational parameters on solar photocatlytic degradation of an azo dye (Direct Yellow 4) using activated carbon-loaded zinc oxide," Materials Science in Semiconductor Processing, vol. 16, pp. 1046–1051, 2013.

202. H.-Y. Zhu, J. Yao, R. Jiang, Y.-Q. Fu, Y.-H. Wu, and G.-M. Zeng, "Enchanced decolarization of azo dye solution by cadmium sulfide.multi-walled carbon nanotubes/polymer composite in combination with hydrogen peroxide under simulated solar light irradiation," Ceramics International, vol. 40, pp. 3769–3777, 2014.

203. N. A. S. Al-Areqi, A. S. N. Al-Kamali, K. A. S. Ghaleb, A. Al_ Alas, and K. Al-Mureish, "Influence of phase stabilization and perovskite vanadate oxygen vacancies of the BINIVOX catalyst on photocatalytic degradation of azo dye under visible light irradiation," Radiation Effects & Defect Solids, vol. 169, pp. 117–128, 2014.

204. C. Andriantsiferana, E. F. Mohamed, and H. Delmas, "Photocatalytic degradation of an azo-dye on TiO_2/activated carbon composite material," Environmental Technologies, vol. 35, pp. 355–363, 2014.

205. J. H. Shariffuddin, M. I. Jones, and D. A. Patterson, "Greener photocatalysts: hydroxyapatite derived from waste mussel shells for the photocatalytic degradation of model azo dye wastewater," Chemical Engineering Research and Design, vol. 91, pp. 1693–1704, 2013.

206. B. Subash, A. Senthilraja, P. Dhatshanamurthi, M. Swaminthan, and M. Shanti, "Solar active photocatalyst for effctive degradation of RR 120 with dye sensitized mechanism,"Spectrochimica Acta, vol. 115, pp. 175–182, 2013.

207. M. H. Habibi and E. Askari, "Spectrophotometric studies of photo-induced degradation of Tertrodirect Light Blue (TLB) using a nanostructure zinc zirconate composite,"Journal of Industrial and Engineering Chemistry, vol. 19, pp. 1400–1405, 2013.

208. H. Liu, G. Li, J. Qu, and H. Liu, "Degradation of azo dye Acid Orange 7 in water by FeO/granular activated carbon system in the presence of ultrasound," Journal of Hazardous Materials, vol. 144, no. 1-2, pp. 180–186, 2007·

209. G. Buitrón, M. Quezada, and G. Moreno, "Aerobic degradation of the azo dye acid red 151 in a sequencing batch biofilter," Bioresource Technology, vol. 92, no. 2, pp. 143–149, 2004. ·

210. S. M. A. G. Ulson de Souza, E. Forgiarini, and A. A. Ulson de Souza, "Toxicity of textile dyes and their degradation by the

enzyme horseradish peroxidase (HRP)," Journal of Hazardous Materials, vol. 147, no. 3, pp. 1073–1078, 2007·

211. W. Zhai, G. Li, P. Yu, L. Yang, and L. Mao, "Silver phosphate/ carbon nanotube-stabilized pickering emulsion for highly efficient photocatalysis," The Journal of Physical Chemistry, vol. 117, pp. 15183–15191, 2013.

212. K. Ullah, Z.-D. Meng, S. Ye, L. Zhu, and W.-C. Oh, "Synthesis and characterization of novel PbS-graphene/TiO$_2$ composite with enchanced photocatalytic activity," Journal of Industrial and Engineering Chemistry, vol. 20, no. 3, pp. 1035–1042, 2014. ·

213. M. Cheng, M. Zhu, Y. Du, and P. Yang, "Enchanced photocatalytic hydrogen evolution based on efficient electron transfer in triphenylaminebased dye functionalized Au@Pt bimetallic core/ shell nanocomposite," International Journal of Hydrogen Energy, vol. 38, pp. 8631–8638, 2013.

214. T. Soltani and M. H. Entezari, "Photolysis and photocatalysis of methylene blue by ferrite bismuth nanoparticles under sunlight irradiation," Journal of Molecular Catalysis A, vol. 377, pp. 197–203, 2013.

215. H.-Y. Sun, C.-B. Liu, Y. Cong, M.-H. Yu, H.-Y. Bai, and G.-B. Che, "New photocatalyst for the degradation of organic dyes based on [Co$_2$(1,4-BDC)(NCP)$_2$]$_n$ · 4nH$_2$O,"Inorganic Chemistry Communications, vol. 35, pp. 130–134, 2013. ·

216. R. Comparelli, E. Fanizza, M. L. Curri, P. D. Cozzoli, G. Mascolo, and A. Agostiano, "UV-induced photocatalytic degradation of azo dyes by organic-capped ZnO nanocrystals immobilized onto substrates," Applied Catalysis B, vol. 60, no. 1-2, pp. 1–11, 2005.

217. S. K. Sharma, H. Bhunia, and P. K. Bajpai, "TiO$_2$-assisted photocatlytic degradation of diazo dye reactive red 120: decolarization kinetics and mineralization investigations,"Journal of Advanced Oxidation Technologies, vol. 16, pp. 306–313, 2013.

218. H. Aliyan, R. Fazaeli, and R. Jalilian, "Fe$_3$O$_4$@mesoporous SBA-15: a magnetically recoverable catayst for photodegradation of malachite green," Applied Surface Science, vol. 276, pp. 147–153, 2013.

219. A. Khanna and V. K. Shetty, "Solar light induced photocatalystic degradation of Reactive Blue 220 (RB 220) dye with highly

efficient Ag@TiO$_2$ core-shell nanoparticles: a comparision with UV photocatalysis," Solar Energy, vol. 99, pp. 67–76, 2014.

220. M.-C. Chang, H.-Y. Shu, T.-H. Tseng, and H.-W. Hsu, "Supported zinc oxide photocatalyst for decolarization and mineralization of orange G dye wastewater under UV 365 irradiation," International Journal of Photoenergy, vol. 2013, Article ID 595031, 12 pages, 2013. ·

221. K. Zhao, Z. Wu, R. Tang, and Y. Jiang, "Preparation of highly visible-light photocatalytic active N-doped TiO$_2$ microcuboids," Journal of the Korean Chemical Society, vol. 57, pp. 489–492, 2013.

222. P. Xiong, L. Wang, X. Sun, B. Xu, and X. Wang, "Ternary titania-cobalt ferrite-polyaniline nanocomposite: a magnetically recycleable hybrid for adsorption and photodegradation of dyes under visible light," Industrial & Engineering Chemistry Research, vol. 52, pp. 10105–10113, 2013.

223. Y.-J. Xu, Y. Zhuang, and X. Fu, "New insight for enhanced photocatalytic activity of TiO$_2$ by doping carbon nanotubes: a case study on degradation of benzene and methyl orange," Journal of Physical Chemistry C, vol. 114, no. 6, pp. 2669–2676, 2010.

224. Y. Zhang, J. Wan, and Y. Ke, "A novel approach of preparing TiO$_2$ films at low temperature and its application in photocatalytic degradation of methyl orange," Journal of Hazardous Materials, vol. 177, no. 1–3, pp. 750–754, 2010. ·

225. M. R. Xu, G. Ni, and F. Zhao, "Preparation, characterization and photocatalytic properties of Cu, P-codoped TiO$_2$ nanoparticles," Advanced Materials Research, vol. 239-242, pp. 2562–2565, 2011·

226. B. Leena, B. Mohit, and K. S. Mohan, "Photocatalysis of giemsa dye: an approach towards biotechnology laboratory effluent treatment," Journal of Environmental & Analytical Toxicology, vol. 113, pp. 1–10, 2011.

227. S. G. Abuabara, L. G. C. Rego, and V. S. Batista, "Influence of thermal fluctuations on interfacial electron transfer in functionalized TiO$_2$ semiconductors," Journal of the American Chemical Society, vol. 127, no. 51, pp. 18234–18242, 2005.

228. I. H. Laura, G. Robert, J. B. Jesse, et al., "Spectral characteristics and photosensitization of TiO_2 nanoparticles in reverse micelles by perylenes," The Journal of Physical Chemistry B, vol. 117, pp. 4568–4581, 2013.

229. Y. Zhang, Z.-R. Tang, X. Fu, and Y.-J. Xu, "TiO_2-graphene nanocomposites for gas-phase photocatalytic degradation of volatile aromatic pollutant: is TiO_2-graphene truly different from other TiO_2-carbon composite materials?" ACS Nano, vol. 4, no. 12, pp. 7303–7314, 2010.

230. I. Arslan, I. A. Balcioğlu, and D. W. Bahnemann, "Advanced chemical oxidation of reactive dyes in simulated dyehouse effluents by ferrioxalate-Fenton/UV-A and TiO_2/UV-A processes," Dyes & Pigments, vol. 47, no. 3, pp. 207–218, 2000. · ·

231. I. A. Alaton, I. A. Balcioglu, and D. W. Bahnemann, "Advanced oxidation of a reactive dyebath effluent: comparison of O_3, H_2O_2/UV-C and TiO_2/UV-A processes," Water Research, vol. 36, no. 5, pp. 1143–1154, 2002.

232. G. Li, J. Qu, X. Zhang, and J. Ge, "Electrochemically assisted photocatalytic degradation of Acid Orange 7 with -PbO_2 electrodes modified by TiO_2," Water Research, vol. 40, no. 2, pp. 213–220, 2006.

233. I. Arslan, I. A. Balcioglu, and D. W. Bahnemann, "Heterogeneous photocatalytic treatment of simulated dyehouse effluents using novel TiO_2-photocatalysts," Applied Catalysis B, vol. 26, no. 3, pp. 193–206, 2000. ·

234. Z.-X. Lu, L. Zhou, Z.-L. Zhang et al., "Cell damage induced by photocatalysis of TiO_2 thin films," Langmuir, vol. 19, no. 21, pp. 8765–8768, 2003. ·

235. H. Yu, S. Chen, X. Quan, H. Zhao, and Y. Zhang, "Fabrication of a TiO_2-BDD heterojunction and its application as a photocatalyst for the simultaneous oxidation of an azo dye and reduction of Cr(VI)," Environmental Science and Technology, vol. 42, no. 10, pp. 3791–3796, 2008. ·

236. Z. Zainal, L. K. Hui, M. Z. Hussein, A. H. Abdullah, and I. K. R. Hamadneh, "Characterization of TiO_2-Chitosan/Glass photocatalyst for the removal of a monoazo dye via photodegradation-adsorption process," Journal of Hazardous Materials, vol. 164, no. 1, pp. 138–145, 2009. ·

237. S. U. M. Khan, M. Al-Shahry, and W. B. Ingler Jr., "Efficient photochemical water splitting by a chemically modified n-TiO_2," Science, vol. 297, no. 5590, pp. 2243–2245, 2002. ·

238. F. Dong, S. Guo, H. Wang, X. Li, and Z. Wu, "Enhancement of the visible light photocatalytic activity of C-doped TiO_2 nanomaterials prepared by a green synthetic approach," Journal of Physical Chemistry C, vol. 115, no. 27, pp. 13285–13292, 2011.

239. P. C. Maness, S. Smolinski, D. M. Blake, Z. Huang, E. J. Wolfrum, and W. A. Jacoby, "Bactericidal activity of photocatalytic TiO_2 reaction: toward an understanding of its killing mechanism," Applied and Environmental Microbiology, vol. 65, no. 9, pp. 4094–4098, 2009.

240. C. Shifu, L. Xuqiang, L. Yunzhang, and C. Gengyu, "The preparation of nitrogen-doped $TiO_{2-x}N_x$ photocatalyst coated on hollow glass microbeads," Applied Surface Science, vol. 253, no. 6, pp. 3077–3082, 2007.

241. D. Chen, D. Yang, Q. Wang, and Z. Jiang, "Effects of boron doping on photocatalytic activity and microstructure of titanium dioxide nanoparticles," Industrial and Engineering Chemistry Research, vol. 45, no. 12, pp. 4110–4116, 2006. · ·

242. W. Li, Y. Bai, C. Liu et al., "Highly thermal stable and highly crystalline anatase TiO_2 for photocatalysis," Environmental Science and Technology, vol. 43, no. 14, pp. 5423–5428, 2009.

243. D. B. Ingram and S. Linic, "Water splitting on composite plasmonic-metal/semiconductor photoelectrodes: evidence for selective plasmon-induced formation of charge carriers near the semiconductor surface," Journal of the American Chemical Society, vol. 133, no. 14, pp. 5202–5205, 2011.

244. J.-H. Sun, S.-Y. Dong, Y.-K. Wang, and S.-P. Sun, "Preparation and photocatalytic property of a novel dumbbell-shaped ZnO microcrystal photocatalyst," Journal of Hazardous Materials, vol. 172, no. 2-3, pp. 1520–1526, 2009. · ·

245. W. Xie, Y. Li, W. Sun, J. Huang, H. Xie, and X. Zhao, "Surface modification of ZnO with Ag improves its photocatalytic efficiency and photostability," Journal of Photochemistry and Photobiology A, vol. 216, no. 2-4, pp. 149–155, 2010

246. W. Chen, W. Lu, Y. Yao, and M. Xu, "Highly efficient decomposition of organic dyes by aqueous-fiber phase transfer and in situ catalytic oxidation using fiber-supported cobalt phthalocyanine," Environmental Science and Technology, vol. 41, no. 17, pp. 6240–6245, 2007

247. A. C. Lucilha, C. E. Bonancêa, W. J. Barreto, and K. Takashima, "Adsorption of the diazo dye Direct Red 23 onto a zinc oxide surface: a spectroscopic study,"Spectrochimica Acta Part A, vol. 75, no. 1, pp. 389–393, 2010.

248. L. S. Andrade, L. A. M. Ruotolo, R. C. Rocha-Filho et al., "On the performance of Fe and Fe,F doped Ti-Pt/PbO$_2$ electrodes in the electrooxidation of the Blue Reactive 19 dye in simulated textile wastewater," Chemosphere, vol. 66, no. 11, pp. 2035–2043, 2007·

249. M. Nasr-Esfahani and M. H. Habibi, "Silver doped TiO$_2$ nanostructure composite photocatalyst film synthesized by sol-gel spin and dip coating technique on glass,"International Journal of Photoenergy, vol. 2008, Article ID 628713, 11 pages, 2008. ·

250. L. Björnsson, P. Hugenholtz, G. W. Tyson, and L. L. Blackall, "Filamentous Chloroflexi (green non-sulfur bacteria) are abundant in wastewater treatment processes with biological nutrient removal," Microbiology, vol. 148, no. 8, pp. 2309–2318, 2002.

251. J. J. Plumb, J. Bell, and D. C. Stuckey, "Microbial populations associated with treatment of an industrial dye effluent in an anaerobic baffled reactor," Applied and Environmental Microbiology, vol. 67, no. 7, pp. 3226–3235, 2001. ·

252. T. Ito, K. Sugita, and S. Okabe, "Isolation, characterization, and in situ detection of a novel chemolithoautotrophic sulfur-oxidizing bacterium in wastewater biofilms growing under microaerophilic conditions," Applied and Environmental Microbiology, vol. 70, no. 5, pp. 3122–3129, 2004.

253. S. M. Burkinshaw and G. W. Collins, "Aftertreatment to reduce the washdown of leuco sulphur dyes on cotton during repeated washing," Journal of the Society of Dyers and Colourists, vol. 114, no. 5-6, pp. 165–168, 1998.

254. R. S. Shraddha, S. Simran, K. Mohit, and K. Ajay, "Laccase: microbial sources, production, purification, and potential

biotechnological applications," Enzyme Research, vol. 2011, Article ID 217861, 11 pages, 2011

255. M. Alvaro, E. Carbonell, M. Esplá, and H. Garcia, "Iron phthalocyanine supported on silica or encapsulated inside zeolite Y as solid photocatalysts for the degradation of phenols and sulfur heterocycles," Applied Catalysis B, vol. 57, no. 1, pp. 37–42, 2005.

256. S.-L. Chen, X.-J. Huang, and Z.-K. Xu, "Functionalization of cellulose nanofiber mats with phthalocyanine for decoloration of reactive dye wastewater," Cellulose, vol. 18, no. 5, pp. 1295–1303, 2011·

257. K. K. Kim, C. S. Lee, R. M. Kroppenstedt, E. Stackebrandt, and S. T. Lee, "Gordonia sihwensis sp. nov., a novel nitrate-reducing bacterium isolated from a wastewater-treatment bioreactor," International Journal of Systematic and Evolutionary Microbiology, vol. 53, no. 5, pp. 1427–1433, 2003

258. L. Zhou, W. Guo, G. Xie, and J. Feng, "Photocatalytic degradation of reactive brilliant red X-3B over BiOI under visible light irradiation," Desalination & Water Treatment, vol. 51, pp. 6517–6525, 2013.

259. D. C. Xu, Z.-W. Lian, M.-L. Fu, B. Yuan, J.-W. Shi, and H.-J. Cui, "Advanced near-infrared-driven photocatalyst: fabrication, characterization and photocatalytic performance of -NaYF$_4$: Yb^{3+}, Tm^{3+}@TiO$_2$ core@ shell microcrystals," Applied Catalysis B, vol. 142, pp. 377–386, 2013.

260. S. Song, L. Xu, Z. He, J. Chen, X. Xiao, and B. Yan, "Mechanism of the photocatalytic degradation of C.I. reactive black 5 at pH 12.0 using SrTiO$_3$/CeO$_2$ as the catalyst,"Environmental Science and Technology, vol. 41, no. 16, pp. 5846–5853, 2007

261. C. Pan and Y. Zhu, "New type of BiPO$_4$ Oxy-acid salt photocatalyst with high photocatalytic activity on degradation of dye," Environmental Science and Technology, vol. 44, no. 14, pp. 5570–5574, 2010.

262. A. Furube, T. Shiozawa, A. Ishikawa, A. Wada, K. Domen, and C. Hirose, "Femtosecond transient absorption spectroscopy on photocatalysts: K$_4$Nb$_6$O$_{17}$ and Ru(bpy)$_3$ $^{2+}$-intercalated K$_4$Nb$_6$O$_{17}$ thin films," Journal of Physical Chemistry B, vol. 106, no. 12, pp. 3065–3072, 2002.

263. U. Ruh, S. Hongqi, W. Shaobin, M. A. Hua, and O. T. Moses, "Wet-chemical synthesis of $InTaO_4$ for photocatalytic decomposition of organic contaminants in air and water with UV-vis light," Industrial & Engineering Chemistry Research, vol. 51, pp. 1563–1569, 2011.

264. V. Gunasekar, B. Divya, K. Brinda, J. Vijakrishnan, V. Ponnusami, and K. S. Rajan, "Enzyme mediated synthesis of Ag-TiO_2 photocatalyst for visible light degradation of reactive dye from aqueous solution," Journal of Sol-Gel Science and Technology, vol. 68, pp. 60–66, 2013.

265. X. Liu, J. Xing, J. Qiu, and X. Sun, "Preparation and characterization of visible light-driven praseodymium-doped mesoporous titania coated magnetite photocatalyst," Indian Journal of Chemistry, vol. 52, pp. 1257–1262, 2013.

266. Y. Y. Wang, H. Xie, W. Zhang, Y. B. Tang, and F. Y. Chen, "Preparation and photocatalytic activity of Fe-Ce-N tri-doped TiO_2 photocatalyst," Advanced Materials Research, vol. 750, pp. 1276–1282, 2013.

267. L. Liu, H. Bai, J. Liu, and D. D. Sun, "Multifuntional graphene oxide-TiO_2-Ag nanocomposites for high performance water disinfection and decontamination under solar irradiation," Journal of Hazardous Materials, vol. 261, pp. 214–223, 2013.

268. C. T. Nam, W.-D. Yang, and L. M. Duc, "Study on photocatalysis of TiO_2 nanotubes prepared by methanol-thermal synthesis at low temperature," Bulletin of Materials Science, vol. 36, pp. 779–778, 2013.

269. X. Lin, D. Fu, L. Hao, and Z. Ding, "Synthesis and enchanced visible-light responsive of C, N, S-tridoped TiO_2 hollow spheres," Journal of Environmental Sciences, vol. 25, pp. 2150–2156, 2013.

270. L.-X. Zhu, Z.-H. Zhao, X.-Y. Yue, and J.-M. Fan, "One-pot hydrothermal synthesis of Ag@Ag2S modified porous TiO_2 and its photocatalytic and antimicrobial properties," Journal of Molecular Catalysis, vol. 27, pp. 467–473, 2013.

271. C. Namasivayam, R. Jeyakumar, and R. T. Yamuna, "Dye removal from wastewater by adsorption on "waste" Fe(III)/Cr(III) hydroxide," Waste Management, vol. 14, no. 7, pp. 643–648, 1994.

272. A. Jain, K. P. Raven, and R. H. Loeppert, "Arsenite and arsenate adsorption on ferrihydrite: surface charge reduction and net OH-release stoichiometry,"Environmental Science and Technology, vol. 33, no. 8, pp. 1179–1184, 1999. · ·

273. A. Gürses, M. Yalçin, and C. Do ar, "Electrocoagulation of some reactive dyes: a statistical investigation of some electrochemical variables," Waste Management, vol. 22, no. 5, pp. 491–499, 2002. · ·

274. H. Lachheb, E. Puzenat, A. Houas et al., "Photocatalytic degradation of various types of dyes (Alizarin S, Crocein Orange G, Methyl Red, Congo Red, Methylene Blue) in water by UV-irradiated titania," Applied Catalysis B, vol. 39, no. 1, pp. 75–90, 2002.

275. C. Guillard, H. Lachheb, A. Houas, M. Ksibi, E. Elaloui, and J.-M. Herrmann, "Influence of chemical structure of dyes, of pH and of inorganic salts on their photocatalytic degradation by TiO_2 comparison of the efficiency of powder and supported TiO_2,"Journal of Photochemistry and Photobiology A, vol. 158, no. 1, pp. 27–36, 2003

276. F. Han, V. S. R. Kambala, M. Srinivasan, D. Rajarathnam, and R. Naidu, "Tailored titanium dioxide photocatalysts for the degradation of organic dyes in wastewater treatment: a review," Applied Catalysis A, vol. 359, no. 1-2, pp. 25–40, 2009·

277. C.-Y. Kuo, "Prevenient dye-degradation mechanisms using UV/ TiO_2/carbon nanotubes process," Journal of Hazardous Materials, vol. 163, no. 1, pp. 239–244, 2009

278. M. Alvaro, E. Carbonell, and H. Garcia, "Photocatalytic degradation of sulphur-containing aromatic compounds in the presence of zeolite-bound 2,4,6-triphenylpyrylium and 2,4,6-triphenylthiapyrylium," Applied Catalysis B, vol. 51, no. 3, pp. 195–202, 2004. ·

279. M. Alvaro, E. Carbonell, H. Garcia, C. Lamaza, and M. Narayana Pillai, "Ship-in-a-bottle synthesis of 2,4,6-triphenylthiapyrylium cations encapsulated in zeolites Y and beta: a novel robust photocatalyst," Photochemical and Photobiological Sciences, vol. 3, no. 2, pp. 189–193, 2004

280. H.-Y. Xu, W.-C. Liu, J. Shi, H. Zhao, and S.-Y. Qi, "Photocatalytic discoloration of Methyl Orange by anatase/schorl composite:

optimization using response surface method," Environmental Science and Pollution Research, vol. 21, no. 2, pp. 1582–1591, 2014

281. J. Cao, C. Zhou, H. Lin, B. Xu, and S. Chen, "Direct hydroysis preparation of plate-like BiOI and their visible light photocatalytic activity for contaminant removal," Materials Letters, vol. 109, pp. 74–77, 2013.

282. S. Sharma, R. Ameta, R. K. Malkani, and S. C. Ameta, "Use of semiconducting bismuth sulfide as a photocatalyst for degradation of Rose Bengal," Macedonian Journal of Chemistry and Chemical Engineering, vol. 30, no. 2, pp. 229–234, 2011.

283. Q. Yan, J. Wang, X. Han, and Z. Liu, "Soft-chemical method for fabrication of $SnO-TiO_2$ nanocomposites with enchanced photocatalytic activity," Journal of Materials Research, vol. 28, pp. 1862–1869, 2013.

284. J. Liu, Q. Yang, W. Yang, M. Li, and Y. Song, "Aquatic plant inspired hierarchical artificial leaves for highly efficient photocatalysis," Journal of Materials Chemistry A, vol. 26, pp. 7760–7766, 2013.

285. A. Y. Stepanov, L. V. Sotnikova, A. A. Vladimirov, D. V. Dyagilev, and T. A. Larichev, "The synthesis and investigation of crystallographic and adsorption properties of TiO_2 powders," Advanced Materials Research, vol. 704, pp. 92–97, 2013.

286. S. Moradi, P. Aberoomand-Azar, S. Raeis-Farshid, S. Abedini-Khorrami, and M. H. Givianrad, "Synthesis and characterzation of $Al-TiO_2/ZnO$ and $Fe-TiO_2/ZnO$ photocatalyst and their photocatalytic behaviour," Asian Journal of Chemistry, vol. 25, pp. 6635–6638, 2013.

287. W.-J. Yoo and S. Kobayashi, "Hydrophosphinylation of unactivated alkenes with secondary phosphine oxides under visible-light photocatalysis," Green Chemistry, vol. 15, pp. 1844–1848, 2013.

288. Y. Zhang, Y. Zhang, and J. Tan, "Novel magnetically separable AgCl/iron oxide composites with enchanced photocatalytic activity driven by visible light," Journal of Alloys and Compounds, vol. 574, pp. 383–390, 2013.

289. H.-N. Cui, J.-Y. Wang, M.-Q. Hu et al., "Efficient photo-driven hydrogen evolution by binuclear nickle catalysts of different coordination in noble-metal-free systems," Dalton Transactions, vol. 42, pp. 8684–8691, 2013.

290. Q. Zhang, C. Tian, A. Wu, Y. Hong, M. Li, and H. Fu, "In situ oxidation of Ag/ZnO by bromine water to prepare ternary Ag-AgBr/ZnO sunlight-derived photocatalyst,"Journal of Alloys and Compounds, vol. 563, pp. 269–273, 2013.

291. L. Li, X. Liu, Y. Zhang et al., "Visible-light photochemical activity of heterostructured core-shell materials composed of selected ternary titanates and ferrites coated by TiO_2,"ACS Applied Materials & Interfaces, vol. 5, pp. 5064–5071, 2013.

292. P. Jiang, D. Ren, D. He, W. Fu, J. Wang, and M. Gu, "An easily sedimentable and effective TiO_2 photocatalyst for removal of dyes in water," Separation & Purification Technology, vol. 122, pp. 128–132, 2014.

293. P. Guo, L. T. Meng, and C. H. Wang, "Core-shell WO_3/TiO_2 nanorod heterostructures for solar light photocatalysis," Advanced Materials Research, vol. 850, pp. 78–81, 2014.

294. K. Ullah, S. Ye, L. Zhu, Z.-D. Meng, S. Sarkar, and W.-C. Oh, "Microwave assisted synthesis of a noble metal-graphene hybrid photocatalyst for high efficient decomposition of organic dyes under visible light," Materials Science and Engineering B, vol. 180, pp. 20–26, 2014

295. R. Adhikari, G. Gyawali, S. H. Cho, R. Narro-Garcia, T. Sekino, and S. W. Lee, "Fe^{3+}/Yb^{3+} co-doped bismuth molybdote nanosheets upconversion photocatalyst with enchanced photocatalytic activity," Journal of Solid State Chemistry, vol. 209, pp. 74–81, 2014.

296. J. Gamage McEvoy, W. Cui, and Z. Zhang, "Synthesis and characterization of Ag/AgCl-activated carbon composites for enchanced visible light photocatalysis," Applied Catalysis B, vol. 144, pp. 702–712, 2014.

297. M. Shamshi Hassan, T. Amma, and M.-S. Khil, "Synthesis of high aspect ratio $CdTiO_3$ nanofibers via electrospinning: characterization and photocatalytic activity," Ceramics International, vol. 40, pp. 423–427, 2014.

298. C. Karakaya, Y. Türker, and O. Dag, "Molten-salt-assisted self-assembly (MASA)-synthesis of mesoporous metal titanate-titania, metal sulfide-titania, and metal selenide-titania thin films," Advanced Functional Materials, vol. 23, pp. 4002–4010, 2013.

299. X. Cai, Y. Cai, Y. Liu et al., "Photocatalytic degradation properties of $Ni(OH)_2$ nanosheets/ZnO nanorods composites for azo dyes under visible-light irradaiation,"Ceramics International, vol. 40, pp. 57–65, 2014

300. X. Zhang, W. Chen, Z. Lin, and J. Yao, "Preparation and photocatalysis performances of bacterial cellulose/TiO_2 composite membranes doped by rare earth elements," Chinese Journal of Materials Research, vol. 24, no. 5, pp. 540–546, 2010.

301. X. Zhang, W. Chen, Z. Lin, J. Yao, and S. Tan, "Preparation and photocatalysis properties of bacterial cellulose/TiO_2 composite membrane doped with rare earth elements," Synthesis and Reactivity in Inorganic, Metal-Organic and Nano-Metal Chemistry, vol. 41, no. 8, pp. 997–1004, 2011

302. G. Manimegalai, S. Shantha Kumar, and C. Sharma, "Pesticide mineralization in water using silver nanoparticles," International Journal of Chemical Sciences, vol. 9, no. 3, pp. 1463–1471, 2011.

303. A. Yahia Cherif, O. Arous, M. Amara, S. Omeiri, H. Kerdjoudj, and M. Trari, "Synthesis of modified polymer inclusion membranes for photo-electrodeposition of cadmium using polarized electrodes," Journal of Hazardous Materials, vol. 227, pp. 386–393, 2012.

304. J. Taranto, D. Frochot, and P. Pichat, "Photocatalytic air purification: comparative efficacy and pressure drop of a TiO_2-coated thin mesh and a honeycomb monolith at high air velocities using a 0.4 m3 close-loop reactor," Separation & Purification Technology, vol. 67, no. 2, pp. 187–193, 2009.

305. A. R. Khataee and M. B. Kasiri, "Photocatalytic degradation of organic dyes in the presence of nanostructured titanium dioxide: influence of the chemical structure of dyes," Journal of Molecular Catalysis A, vol. 328, no. 1-2, pp. 8–26, 2010.

Chapter 4

Comparative Study of the Photocatalytic Activity of Semiconductor Nanostructures and Their Hybrid Metal Nanocomposites on the Photodegradation of Malathion

Dina Mamdouh Fouad[1] and Mona Bakr Mohamed[2, 3]

[1]Chemistry Department, Faculty of Science, Assiut University, Assiut 71516, Egypt

[2]National Institute of Laser Enhanced Science, Cairo University, Giza, Egypt

[3]NanoTech Egypt for Photoelectronic, Dreamland, October City, Egypt

ABSTRACT

This work is devoted to synthesize different semiconductor nanoparticles and their metal-hybrid nanocomposites such as TiO_2, Au/TiO_2, ZnO, and Au/ZnO. The morphology and crystal structure of the prepared nanomaterials are characterized by the TEM and XRD, respectively. These materials are used as catalysts for the photodegradation of Malathion which is one of the most commonly used pesticides in the developing countries. The degradation of 10 ppm Malathion under ultraviolet (UV) and visible light in the presence of the different synthesized nanocomposites was analyzed with high-performance liquid chromatography (HPLC) and UV-Visible Spectra. A comprehensive study is carried out for the catalytic efficiency of the prepared nanoparticles. Different factors influencing the catalytic photodegradation are investigated, as different light source, surface coverage, and nature of the organic contaminants. The results indicate that hybrid nanocomposite of the semiconductor-metal hybrid serves as a better catalytic system compared with semiconductor nanoparticles themselves.

INTRODUCTION

Malathion, an organophosphorous pesticide with abroad range of target pests, has been widely used in agriculture. Malathion is suspected to cause child leukemia, anemia, and kidney failure and is widely used in developing countries [1, 2]; it can persist in the human body for at least two generations [3, 4]. However, due to its chemical stability and high toxicity, Malathion resists to biodegrade [5]. Therefore, it was important to explore a new methodology for reducing the contamination of water with Malathion. Photocatalysis is considered to be one of the most potential pollution remediation technologies in recent decades [6, 7]. In recent years, semiconductor photocatalysis has become more and more attractive and important since it has a great potential to contribute to such environmental problems. One of the most important aspects of environmental photocatalysis is the selection of the semiconductor material. Semiconductor photocatalyst generates electron and hole pair (e^--h^+) upon irradiation of light energy that could be utilized in initiating oxidation and reduction reactions of the pesticide, respectively. Guillard et al. [8] reported that the number of photons striking the photocatalyst actually controls the rate of the reaction which is an indication that the reaction takes place only on the adsorbed phase of the semiconductor particle. Surface morphology, namely, the particle size and shape is a very important parameter influencing the performance of photocatalyst in photocatalytic oxidation [9]. TiO_2 is one of the most commonly studied photocatalyst for it is easily available, relatively inexpensive, and chemically stable [10, 11]. However, TiO_2 is incompetent due to the wide band gap, it can only be triggered by near UV radiation, the photo generated electron and hole pairs are liable to recombination, leading to low quantum yields [12, 13]. ZnO has received much attention in the degradation of environmental pollutants since it has almost the same band gap energy (3.2 eV) as TiO_2. However, photo-corrosion frequently occurs with illumination of UV light and is considered as one of the main reasons for the decrease of ZnO photocatalytic activity.

Nanosized Gold particle possesses many excellent properties, such as easy reductive preparation, water solubility, high chemical stability, and significant biocompatibility and affinity [14]. Surface plasmon resonance absorption is a unique property of gold which is the ability of showing a strong absorption band in the visible region when the frequency of the electromagnetic field is resonant with the coherent electron motion [15]. Combining the semiconductor and metallic character in the same nanomaterial could enhance the catalytic activity due to the increase of the photo-absorption resulting from the plasmonic effect and also increases the rate of charge separation at the interface.

The prospect of developing new multifunctional nanocomposites of metal gold hybridized with inorganic components has become of great importance [16–18]. Bimetallic colloids are interesting from a number of perspectives, such as their unique electronic, catalytic, and optical properties [19–21]; photodegradation of Malathion has been studied in aqueous solution using an Au-Pd-TiO_2 nanotubes film [22].

In the present work, we tested the suitability of using semiconductor nanocomposites (TiO_2, ZnO) and metal semiconductors nanocomposite (Au/TiO_2, Au/ZnO) for the photodegradation of Malathion. The photocatalytic activities of the different synthesized nanoparticles are compared, and the influence of different parameters is studied, such as surface coverage and light source irradiation. The efficiency of the catalytic activity showed dependence on source of light irradiation, particle size, crystalline, surface coverage of the nanocomposite, and nature of the organic contaminate.

MATERIALS AND METHODS

Chemicals

Malathion (99% HPLC grade) was purchased from Fluka and used as received. Hydrogen tetrachloroaurate trihydrate (HAuCl4.3H$_2$O) (99.9%) was purchased from Sigma-Aldrich, and Polyvinyl pyrrolidone PVP-K30 (Av. Wt. 22000) was purchased from Fluka. Tri-sodium citrate (99%) was purchased from (Sigma-Aldrich), sterile sodium chloride physiological saline (0.9%) (ADWIC). TiO$_2$, ZnO and HPLC grade solvents (purity 99%) such as methanol and ethanol were purchased from Aldrich. High purity water used in the experiments was purified with the milli-Q system. All chemicals were used without any further purification.

Synthesis of Semiconductor Nanoparticles

Synthesis of TiO$_2$ Nanoparticles

The precursor of 5 mL Titanium isopropoxide is added to a mixture of 5 mL of isopropyl alcohol and 3 mL glacial acetic acid dropwise with constant stirring. The prepared particles are separated with centrifuge and dried for further characterization. The particle size and shape is determined using TEM and the crystal structure is determined using XRD. The optical absorption is measured using UV-Visible PerkinElmer Lambda 40 double beam spectrophotometer.

Synthesis of ZnO Nanoparticles

1.48 g (10 mmoL) of Zn(CH$_3$COO)$_2$2H$_2$O and 1.38 g (23.8 mmoL) of NaHCO$_3$ are mixed at room temperature. The mixture is ignited at 300°C for 3 hours. The $Zn(CH_3COO)_2 2H_2O$ is converted to ZnO nanoparticles, while the $NaHCO_3$ is converted to CH_3COONa and eventually washed away with deionized water until the formation of white ZnO nanoparticles. The particle size and shape is determined using TEM, and the crystal structure is determined using XRD.

Synthesis of Gold Nanoparticles by Citrate Method

Spherical gold nanoparticles (GNPs) were prepared in aqueous solution according to a method described by Turkevich. Simply, the method is a chemical reduction of gold ions by sodium citrate in aqueous solution. Sodium citrate serves also as a capping material which prevents aggregation and further growth of the particles. 5 mL of 1% sodium citrate solution was added to 40 mL chloroauric acid (HAuCl4) boiling solution containing 5 mg of gold ions. The solution was boiled for 30 minutes and was then left to cool down to room temperature. The particle size and shape is determined using TEM, and the crystal structure is determined using XRD.

Synthesis Au/TiO_2 and Au/ZnO Nanoparticles

Gold nanoparticles are prepared by citrate method as shown above, and the obtained particles are used as a seed to grow TiO_2 or ZnO nanoshell. The reaction mixture is microwaved for 12 minutes. The particle size and shape is determined using TEM, and the crystal structure is determined using XRD.

Photodegradation of Malathion

Photodegradation radiation rate of Malathion was carried out using Bischof HPLC system (Mainz, Germany), C18, reversed Colum (250 × 4.6 mm), and UV-detector with variable wave length 250 nm was used. Methanol/water (70/30) was used as an eluent at flow rate of 1 mL/min at retention times 12 min. $20 \mu L$ was injected to the HPLC at periodic interval 20 minutes. The adsorption of the pesticides on the nanoparticles resulted in the gradual decrease in the peak area and area percent in comparison to the peak area and area percent of the standard solution of the Malathion. An interaction was carried out between 10 ppm of Malathion (Figure 1) with the different synthesised nanoparticles (TiO_2, ZnO, Au/ZnO, and Au/TiO_2) of three different concentrations (10^{-4}, 3×10^{-5}, 10^{-5} M); the different aliquots were subjected to UV lamp, natural sunlight and investigated at equal time intervals 30 min and analyzed by using Perkin Elemer 240 spectrophotometer. Data acquisition and manipulation were performed using computer-based program.

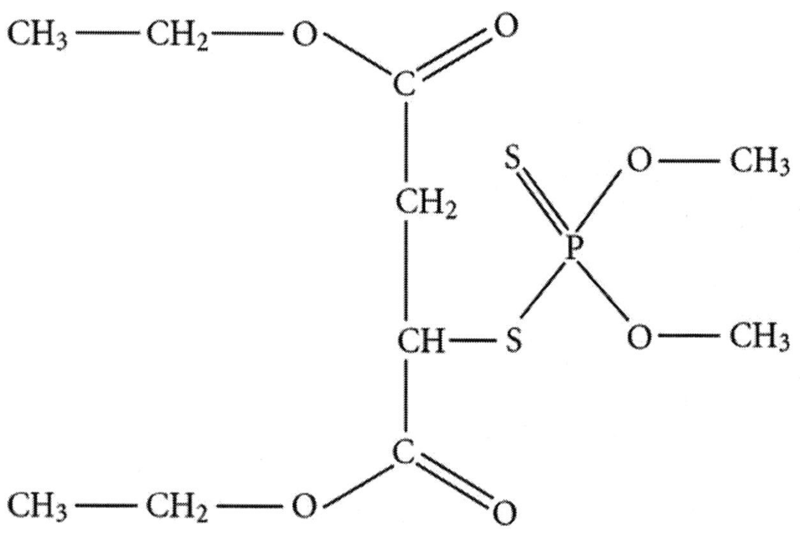

Figure 1: Chemical structure of Malathion.

RESULTS AND DISCUSSION

Characterization of the Synthesized Nanocomposites

Different metal oxide nanoparticles such as TiO_2, ZnO, and their core-shell gold nanocomposites has been synthesized chemically as shown in the experimental part and characterized using absorption spectra,

TEM and XRD. Figures 2, 3, 4, and 5 show the absorption spectra associated with TEM images for all the particles prepared. As shown in Figures 2–5, the prepared nanoparticles have monodispersed size and shapes and their size is less than 100 nm and a band gap absorption in the UV region; accordingly irradiation with UV light only creates (e^--h^+) pair and activates the material to be a photocatalyst. Presence of gold with the nanocomposite enhances its absorption coefficient and increases the photocatalytic activity due to the increase in the charge separation rate. Also the gold-semiconductor composite has absorption at visible region due to the surface plasmon of the gold particle; this means that the photocatalytic activity of gold could be initiated by irradiation with both UV and visible lights.

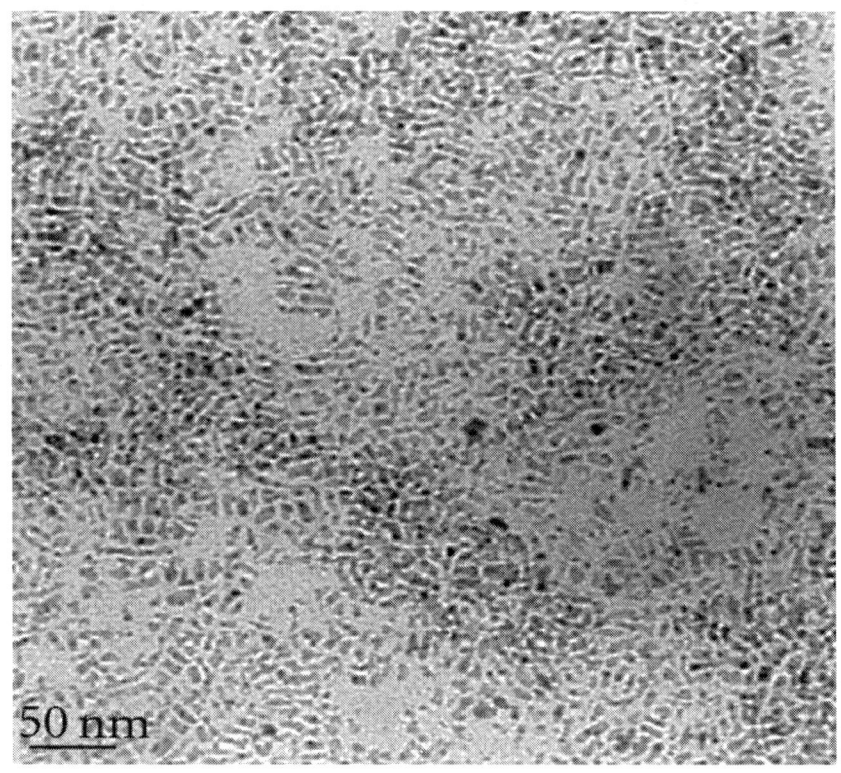

50 nm

(a)

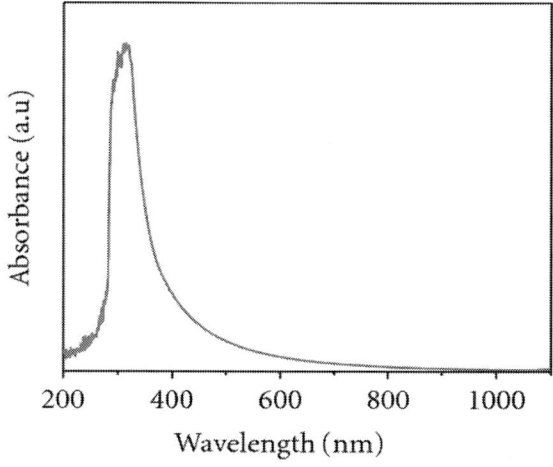

(b)

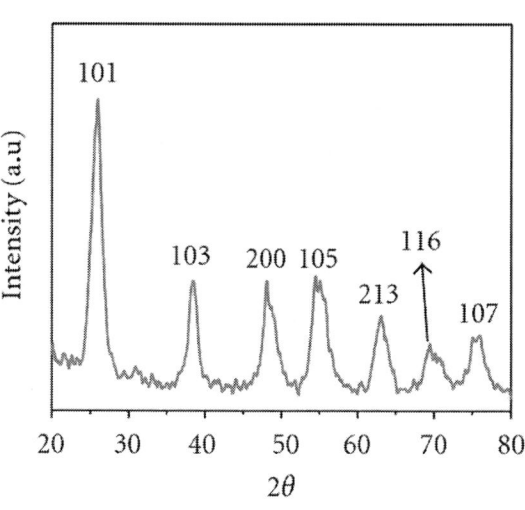

(c)

Figure 2: Shows TEM image (a), UV-Vis absorption spectrum (b), and XRD patterns (c) of prepared TiO_2 nanoparticles.

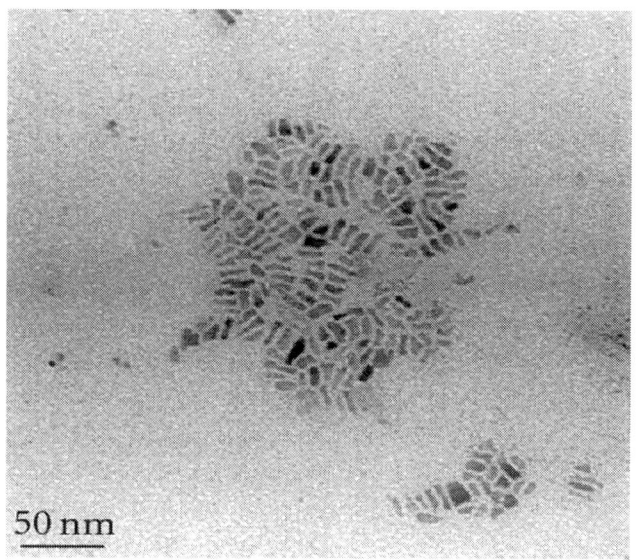

(a)

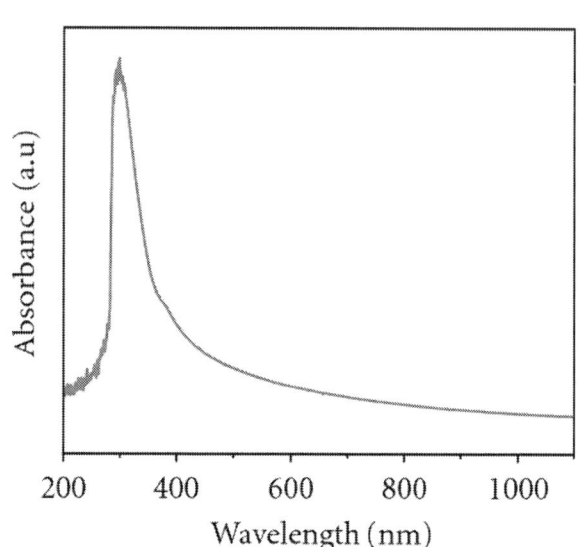

(b)

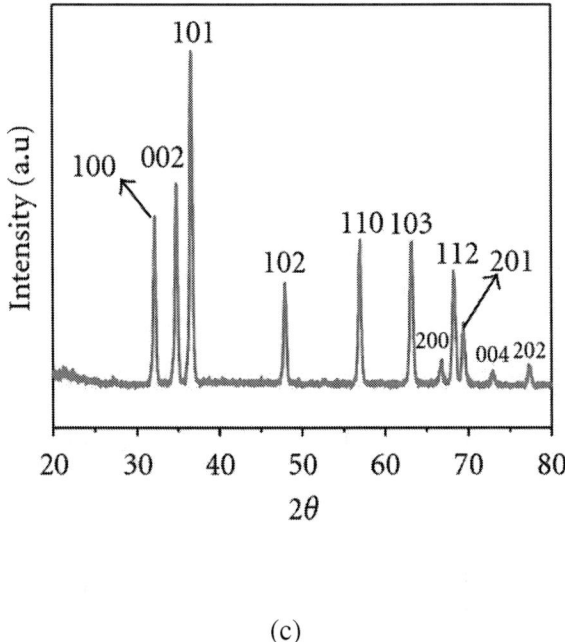

(c)

Figure 3: Shows TEM image (a), UV-Vis absorption spectrum (b), and XRD patterns (c) of prepared ZnO nanoparticles.

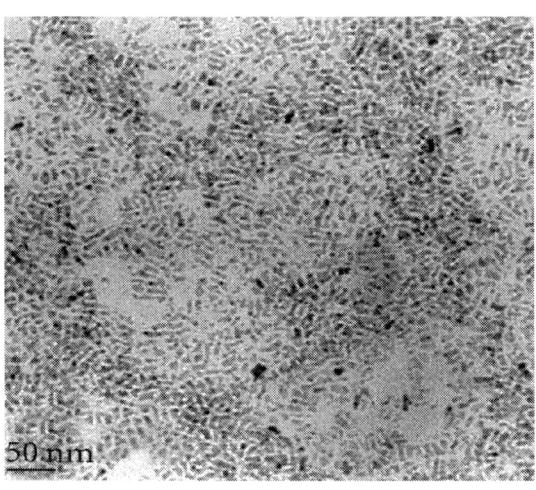

(a)

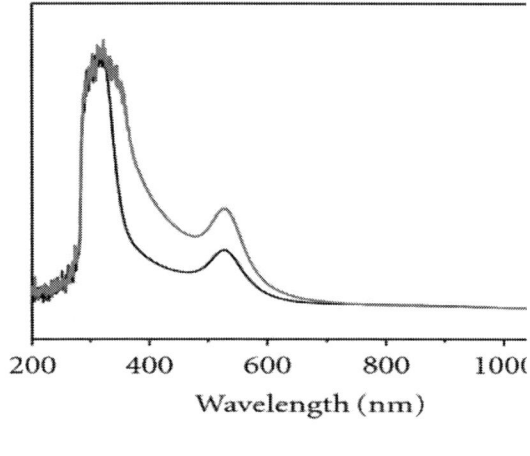

(b)

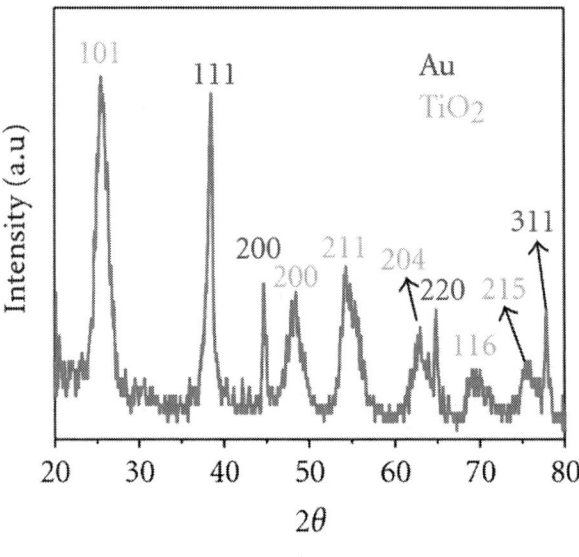

(c)

Figure 4: Shows TEM image (a), UV-Vis absorption spectrum (b), and XRD patterns (c) of prepared Au/TiO$_2$ nanoparticles.

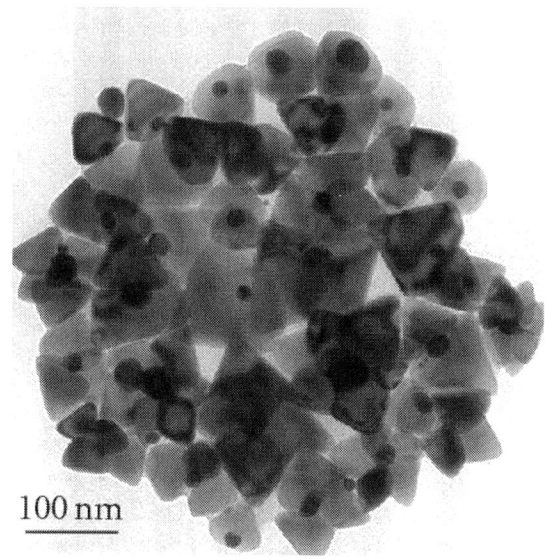

(a)

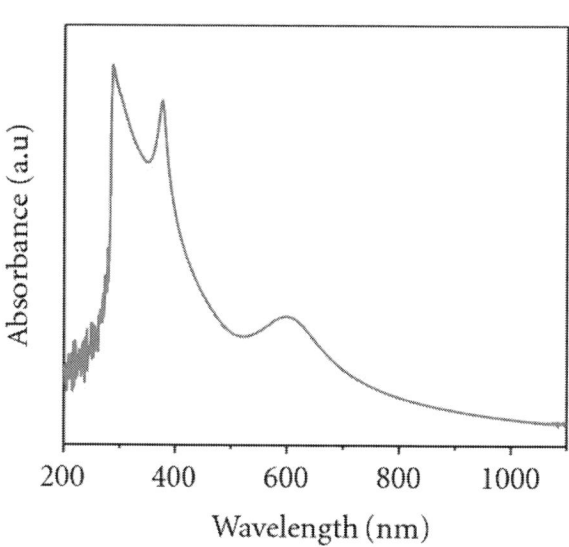

(b)

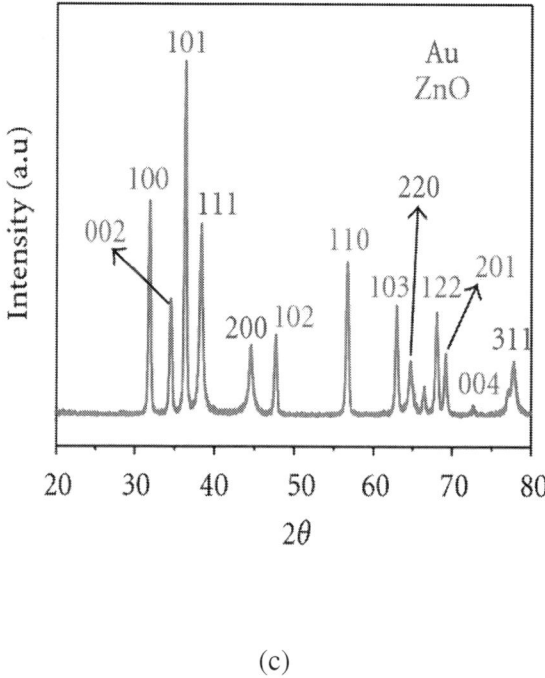

(c)

Figure 5: Shows TEM image (a), UV-Vis absorption spectrum (b), and XRD patterns (c) of prepared Au/ZnO nanoparticles.

Effect of Different Light Sources

Malathion has a characteristic absorption band at 260 nm; however, the rate of photodegradation is followed using the decay of this band. The degradation of Malathion was carried out under the irradiation of light from two different sources, sun light (which is mainly visible light) and UV lamp which emits light around 254 nm (Table 1). It has been reported that the photodegradation rate increases as increasing the light intensity during photocatalytic degradation reaction [22–25]. Our results clearly indicate that UV irradiation causes higher rate of degradation for Malathion than with sun light, due to high intensity of light which is suitable for the excitation of many electrons from the valence band of the metal oxide semiconductor as illustrated in Figure 6. The band gap for TiO_2 and ZnO lies in the UV region, thus, using UV light initiates the excitation of electrons from the conduction band into the valance band due to the formed electron or active oxygen which are responsible for the degradation of the Malathion. Presence of the gold in the photocatalyst permits the photodegradation of Malathion in natural sun light since it is a visible light responsive catalyst having an absorption band around 520 nm; this explains the marked increase in photodegradation of Malathion with Au/ZnO and Au/TiO$_2$. The catalytic activity is markedly enhanced by doping small amounts of metals such as Au which prevents the electron hole recombination and accelerates the photocatalytic degradation with UV light to a greater extent as shown in (Table 1). Figure 7 presents the HPLC chromatograms for Malathion alone, and after the addition of the different nanoparticles, Malathion showed a marked decrease in the peak height and the integration area percentage in addition to the Au/TiO$_2$ and Au/ZnO nanoparticles to Malathion in comparison to the photodegradation using TiO$_2$ and ZnO as presented in (Table 2), which adopts the idea that the presence of gold effectively scavenged the holes, thus competing with the charge recombination which in terms accelerates the photodegradation.

Table 1: Degradation of Malathion after 1 hour on irradiation to UV-lamp and natural sun light

Photocatalysts	Degradation rate %	
	Irradiation to sun light	Irradiation to UV-C lamp
TiO2	6	68
ZnO	5	60
Au/ZnO	62	79
Au/TiO2	67	81

Table 2: HPLC integration area for Malathion on irradiation to UV

Photocatalyst	Integration area (%)	
	At once	After an hour
TiO2	83	40
Au/TiO2	75	19
Au/ZnO	85	27

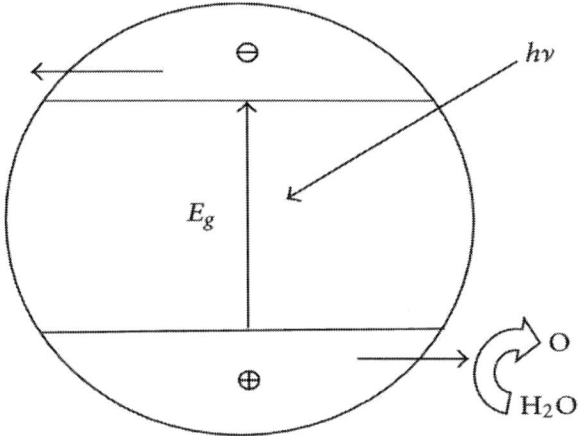

Figure 6: Scheme shows the roles of the Semiconductor metal oxide in photodegradation which creates active oxygen causing decomposition of water pollutants.

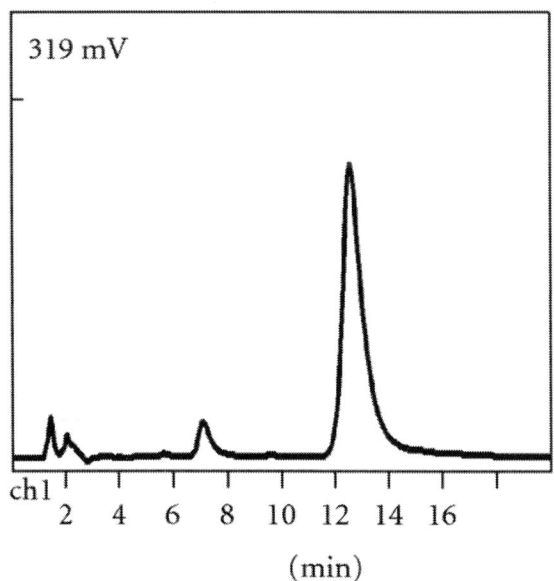

(a)

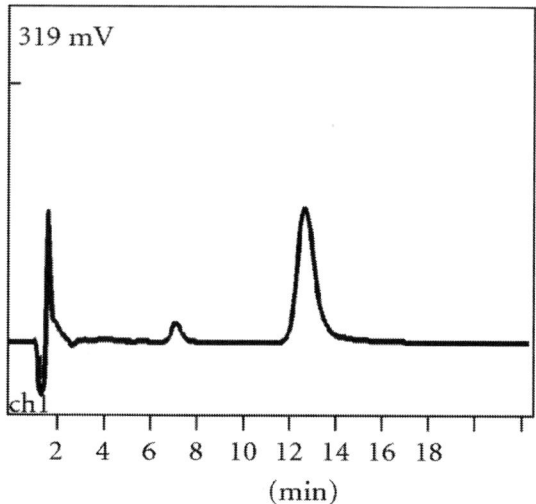

(b)

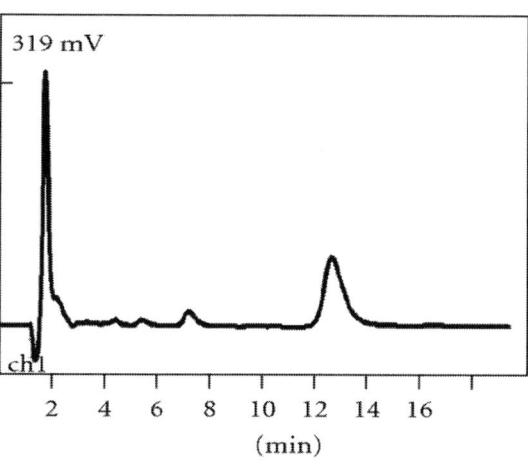

(c)

Figure 7: Shows HPLC chromatogram for Malathion (a), Malathion + Au/ZnO (b), and Malathion Au/TiO$_2$ (c).

Effect of the Nature of Photocatalyst

A direct correlation exists between the removal of the organic pollutant and the surface coverage of TiO_2 photocatalyst [26]. Heterogeneous photocatalytic reactions are known to show a proportional increase in the photodegradation with catalyst loading [27]. Generally, in any given photocatalytic application, the optimum catalyst concentration must be determined, in order to avoid the excess catalyst and ensure the total absorption of the efficient photons [28]. This is because the scattering of an unfavorable light and reduction of light penetration into the solution is observed with excess photocatalyst loading [29]. During photocatalytic oxidation process, the concentration of organic substrate over time is dependent upon photonic efficiency [30]. At high-substrate concentrations, however, the photonic efficiency diminishes and the titanium dioxide surface becomes saturated leading to catalyst deactivation [31]. Different concentrations of the nanoparticles were investigated to select the optimal concentration for efficient photodegradation. As presented in Figure 8, it is concluded that 10^{-4} M is the optimum concentration for efficient degradation of Malathion using the different nanoparticles; about 69% was achieved in 1 hour using TiO_2; at lower concentrations (3×10^{-5}, 10^{-4}) free radical production rate is limited which suppresses the rate of the degradation reaction to (10–20%). The same trend was obtained for the prepared nanoparticles as presented in Figures 9-10; an increase of the degradation percentage was observed on using Au/ZnO and Au/TiO_2 and also a reduction of the radiation rate time, which is due to the presence of the gold which prevents the electron hole recombination and accelerates the photodegradation.

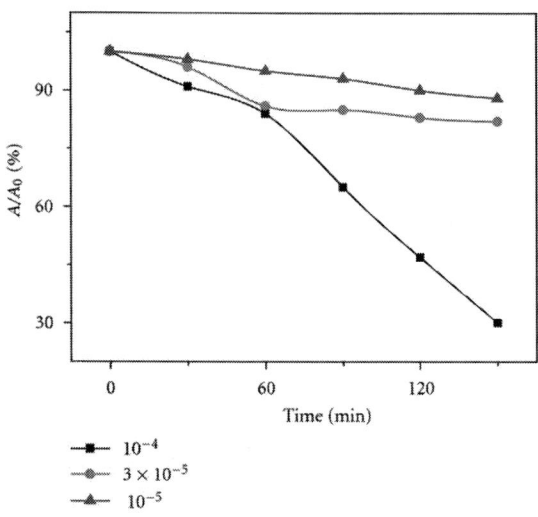

Figure 8: Shows the effect of different concentrations of TiO$_2$ nanoparticle on the degradation of Malathion.

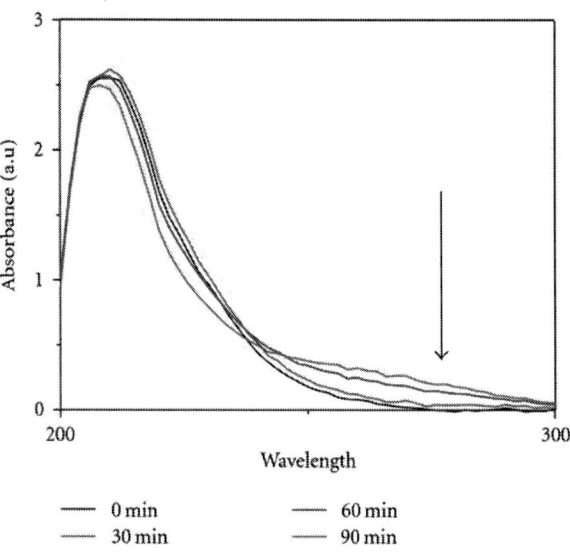

Figure 9: Shows time-dependent degradation of Malathion using TiO$_2$ in UV lamp.

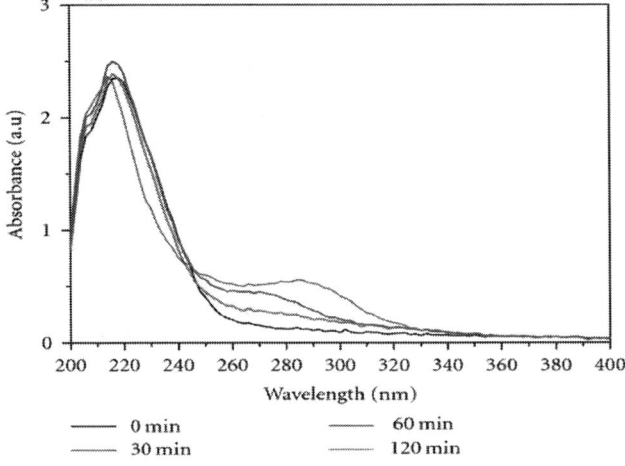

Figure 10: Shows time-dependent degradation of Malathion using Au/ZnO in UV lamp.

Effect of the Nature of the Contaminant

Organic molecules which adhere effectively to the surface of the photocatalyst are more susceptible to direct oxidation [32]. Thus the photocatalytic degradation of aromatics depends on the substituent group. It is reported that nitro-phenol is much stronger adsorbing substrate than phenol and therefore degrades faster [33]. In the degradation of chloroaromatics, Bhatkhande et al. [34] pointed out that monochlorinated phenol degrades faster than di- or tri-chlorinated member. In general, molecules with electron withdrawing groups such as nitrobenzene and benzoic acid were found to adsorb significantly in the dark compared to those with electron donating groups [35]. During photocatalytic oxidation process, the concentration of organic substrate over time is dependent upon photonic efficiency [36]. At high-substrate concentrations, however, the photonic efficiency diminishes and the titanium dioxide surface becomes saturated leading to catalyst deactivation [37]. In the present work, we compared the degradation of two different pesticides as a model for aliphatic (Malathion) and chloroaromatic (chloridazone) by subjecting them to the same concentration of different categories of nanoparticle Au/TiO$_2$ and Au/

ZnO and for same period of time, about 1 hour. As indicated in Figure 11, it is clearly shown that the degradation of chloridazone is faster in comparison to Malathion; a similar trend of degradation was obtained for the both types of nanoparticles since after irradiation to UV lamp for 30 min the degradation of Malathion was about 30%, while in case of chloridazone it was about 50%; this could be explained in terms of the presence of aromatic rings as well as the number and nature of substituent on the ring (like electron donating or electron withdrawing groups) which are known to affect the adsorption and consequently the degradation rate [38].

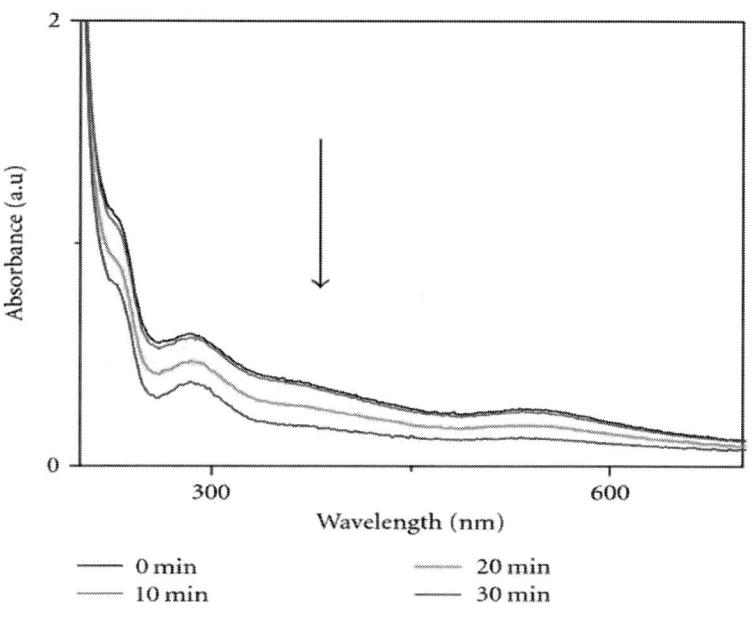

(a)

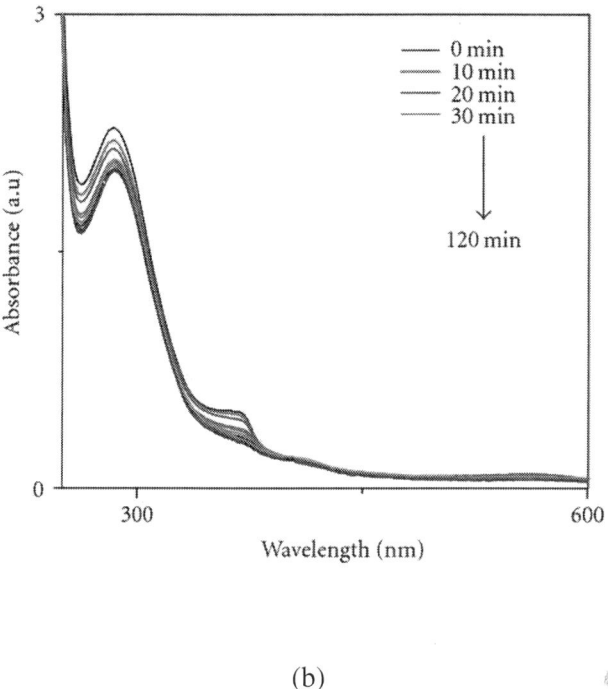

(b)

Figure 11: Shows time-dependent photodegradation of cholridazon in UV lamp after addition of 10^{-4} TiO$_2$/Au (a) and ZnO/Au (b).

CONCLUSIONS

The ability to synthesize multicomponent nanocomposites is important to improve the electronic, optical, and magnetic functionality. According to our study, gold hybrid semiconductor-noble metal nanocrystals not only combine the unique properties of the metal and semiconductors but also generate collective new phenomena based on the intraparticles interaction between the metal and the semiconductor at their interface. The presence of metal-semiconductor interface promotes effective charge separation carrier transfers which subsequently enhance photocatalytic effect. Photodegradation of 10 ppm Malathion was enhanced in the presence of gold-semiconductor nanoparticle; other factors also influenced the degradation rate such as different light sources and nature of the catalyst.

REFERENCES

1. J. L. Adgate, D. B. Barr, C. A. Clayton et al., "Measurement of children's exposure to pesticides: analysis of urinary metabolite levels in a probability-based sample," Environmental Health Perspectives, vol. 109, no. 6, pp. 583–590, 2001.

2. D. Zeljezic and V. Garaj-Vrhovac, "Evaluation of genetic damage in workers employed in pesticide production utilizing the comet assay," Chemosphere, vol. 46, pp. 295–303, 2002.

3. T. Vial, B. Nicolas, and J. Descotes, "Clinical immunotoxicity of pesticides," Journal of Toxicology and Environmental Health, vol. 48, no. 3, pp. 215–229, 1996.

4. A. S. Nair, C. Subramaniam, M. J. Rosemary et al., "Nanoparticles-chemistry, new synthetic approaches, gas phase clustering and novel applications," Pramana Journal of Physics, vol. 65, no. 4, pp. 631–640, 2005.

5. B. Kumari, A. Guha, M. G. Pathak, T. C. Bora, and M. K. Roy, "Experimental biofilm and Its application in malathion degradation," Folia Microbiologica, vol. 43, no. 1, pp. 27–30, 1998.

6. I. Oller, W. Gernjak, M. I. Maldonado, L. A. Pérez-Estrada, J. A. Sánchez-Pérez, and S. Malato, "Solar photocatalytic degradation of some hazardous water-soluble pesticides at pilot-plant scale," Journal of Hazardous Materials, vol. 138, no. 3, pp. 507–517, 2006.

7. A. Corma and H. Garcia, "Zeolite-based photocatalysts," Chemical Communications, vol. 10, no. 13, pp. 1443–1459, 2004.

8. C. Guillard, H. Lachheb, A. Houas, M. Ksibi, E. Elaloui, and J. M. Herrmann, "Influence of chemical structure of dyes, of ph and of inorganic salts on their photocatalytic degradation by TiO_2 comparison of the efficiency of powder and supported TiO_2," Journal of Photochemistry and Photobiology A, vol. 158, no. 1, pp. 27–36, 2003.

9. K. Kogo, H. Yoneyama, and H. Tamura, "Photocatalytic oxidation of cyanide on platinized TiO_2," Journal of Physical Chemistry, vol. 84, no. 13, pp. 1705–1710, 1980.

10. J. Zhao and X. Yang, "Photocatalytic oxidation for indoor air purification: a literature review," Building and Environment, vol. 38, no. 5, pp. 645–654, 2003.

11. F. Zang, J. Zhao, T. Shen, et al., "TiO_2-assisted photode gradation of dye pollutants. II. Adsorption and degradation kinetics of eosin in TiO_2 dispersions under visible light irradiation," Applied Catalysis B, vol. 5, pp. 147–156, 1998.

12. G. Rothenberger, J. Moser, M. Grätzel, N. Serpone, and D. K. Sharma, "Charge carrier trapping and recombination dynamics in small semiconductor particles," Journal of the American Chemical Society, vol. 107, no. 26, pp. 8054–8059, 1985.

13. H. Tributsch, N. Serpone, and E. Pelizzeti, Photocatalysis: Fundamentals and Applications, Wiley, New York, NY, USA, 1989.

14. M. C. Daniel and D. Astruc, "Gold aanoparticles: assembly, supramolecular chemistry, quantum-size-related properties, and applications toward biology, catalysis, and nanotechnology," Chemical Reviews, vol. 104, no. 1, pp. 293–346, 2004.

15. C. F. Bohren and D. R. Huffman, Absorption and Scattering of Light by Small Particles, Wiley, New York, NY, USA, 1983.

16. J. L. Lyon, D. A. Fleming, M. B. Stone, P. Schiffer, and M. E. Williams, "Synthesis of Fe oxide core/Au shell nanoparticles by iterative hydroxylamine seeding," Nano Letters, vol. 4, no. 4, pp. 719–723, 2004.

17. S. J. Cho, J. C. Idrobo, J. Olamit, K. Liu, N. D. Browning, and S. M. Kauzlarich, "Growth mechanisms and oxidation resistance of gold-coated iron nanoparticles," Chemistry of Materials, vol. 17, no. 12, pp. 3181–3186, 2005.

18. D. Caruntu, B. L. Cushing, G. Caruntu, and C. J. O'Connor, "Attachment of gold nanograins onto colloidal magnetite nanocrystals," Chemistry of Materials, vol. 17, no. 13, pp. 3398–3402, 2005.

19. Y. H. Chen and U. Nickel, "Superadditive catalysis of homogeneous redox reactions with mixed silver-gold colloids," Journal of the Chemical Society, Faraday Transactions, vol. 89, no. 14, pp. 2479–2485, 1993.

20. N. Aihara, K. Torigoe, and K. Esumi, "Preparation and characterization of gold and silver nanoparticles in layered laponite suspensions," Langmuir, vol. 14, no. 17, pp. 4945–4949, 1998.

21. M. Michaelis, A. Henglein, and P. Mulvaney, "Composite Pd-Ag particles in aqueous solution," Journal of Physical Chemistry, vol. 98, no. 24, pp. 6212–6215, 1994.

22. M. Muneer, M. Qamar, and D. Bahnemann, "Heterogeneous photocatalysed reaction of two selected pesticide derivatives trichlopyr and daminozid in aqueous suspension of titanium dioxide," Journal of Environmental Management, vol. 80, pp. 99–106, 2006.

23. C. Karunakaran and S. Senthilvelan, "Photocatalysis with ZrO_2: oxidation of aniline," Journal of Molecular Catalysis A: Chemical, vol. 233, no. 1-2, pp. 1–8, 2005.

24. M. Stylidi, D. I. Kondarides, and X. E. Verykios, "Visible light-induced photocatalytic degradation of acid orange 7 in aqueous TiO_2 suspensions TiO_2 suspensions," Applied Catalysis B, vol. 47, no. 3, pp. 189–201, 2004.

25. K. Wilke and H. D. Breuer, "The influence of transition metal doping on the physical and photocatalytic properties of titania," Journal of Photochemistry and Photobiology A, vol. 121, no. 1, pp. 49–53, 1999.

26. J. Araña, J. L. Martínez Nieto, J. A. H. Melián et al., "Photocatalytic degradation of formaldehyde containing wastewater from veterinarian laboratories," Chemosphere, vol. 55, no. 6, pp. 893–904, 2004.

27. A. J. Maira, K. L. Yeung, J. Soria et al., "Gas-phase photo-oxidation of toluene using nanometer-size TiO_2 catalysts," Applied Catalysis B, vol. 29, no. 4, pp. 327–336, 2001.

28. J. Krýsa, M. Keppert, J. Jirkovský, V. Štengl, and J. Šubrt, "The effect of thermal treatment on the properties of TiO_2 photocatalyst," Materials Chemistry and Physics, vol. 86, no. 2-3, pp. 333–339, 2004.

29. M. Saquib and M. Muneer, "TiO_2-mediated photocatalytic degradation of a triphenylmethane dye (gentian violet), in aqueous suspensions," Dyes and Pigments, vol. 56, no. 1, pp. 37–49, 2003.

30. G. Palmisano, M. Addamo, V. Augugliaro et al., "Selectivity of hydroxyl radical in the partial oxidation of aromatic compounds in heterogeneous photocatalysis," Catalysis Today, vol. 122, no. 1-2, pp. 118–127, 2007.

31. D. A. Friesen, L. Morello, J. V. Headley, and C. H. Langford, "Factors influencing relative efficiency in photo-oxidations of organic molecules by Cs3PWO and TiO$_2$ colloidal photocatalysts," Journal of Photochemistry and Photobiology A, vol. 133, no. 3, pp. 213–220, 2000.

32. N. Serpone, "Relative photonic efficiencies and quantum yields in heterogeneous photocatalysis," Journal of Photochemistry and Photobiology A, vol. 104, no. 1-3, pp. 1–12, 1997.

33. U. I. Gaya and A. H. Abdullah, "Heterogeneous photocatalytic degradation of organic contaminants over titanium dioxide: a review of fundamentals, progress and problems," Journal of Photochemistry and Photobiology C: Photochemistry Reviews, vol. 9, no. 1, pp. 1–12, 2008.

34. D. S. Bhatkhande, S. P. Kamble, S. B. Sawant, and V. G. Pangarkar, "Photocatalytic and photochemical degradation of nitrobenzene using artificial ultraviolet light," Chemical Engineering Journal, vol. 102, no. 3, pp. 283–290, 2004.

35. M. Hügül, E. Erçağ, and R. Apak, "Kinetic studies on UV-photodegradation of some chlorophenols using TiO$_2$ catalyst," Journal of Environmental Science and Health A, vol. 37, no. 3, pp. 365–383, 2002.

36. G. Palmisano, M. Addamo, V. Augugliaro et al., "Selectivity of hydroxyl radical in the partial oxidation of aromatic compounds in heterogeneous photocatalysis," Catalysis Today, vol. 122, no. 1-2, pp. 118–127, 2007.

37. D. A. Friesen, L. Morello, J. V. Headley, and C. H. Langford, "Factors influencing relative efficiency in photo-oxidations of organic molecules by Cs3PWO and TiO$_2$ colloidal photocatalysts," Journal of Photochemistry and Photobiology A, vol. 133, no. 3, pp. 213–220, 2000.

38. P. V. Kamat, R. Huehn, and R. Nicolaescu, "A "sense and shoot" approach for photocatalytic degradation of organic contaminants in water," Journal of Physical Chemistry B, vol. 96, no. 4, pp. 788–794, 2007.

Chapter 5

Synthesis, Characterization, and Photocatalytic Activity of N-Doped Zno/Zns Composites

Hongchao Ma, Xiaohong Cheng, Chun Ma, Xiaoli Dong, Xinxin Zhang, Mang Xue, Xiufang Zhang, and Yinghuan Fu

School of Chemistry Engineering and Material, Dalian Polytechnic University, Dalian, Liaoning 116034, China

ABSTRACT

The aim of the present study is to enhance photocatalytic performance of ZnO semiconductor by comodification with doping of nonmetal ions and coupling with another semiconductor. Therefore, we synthesized the N-doped ZnO/ZnS photocatalysts via a simple heat-treatment approach using L-cysteine as N and S source in this work. Anthraquinone dye (reactive brilliant blue KNR) is employed as the model contaminants to evaluate the photocatalytic activity of as-synthesized samples under sunlight illumination. The N-doped ZnO/ZnS synthesized by this method shows better photocatalytic activity as compared to that of pure ZnO. The enhanced photocatalytic activity of the N-doped ZnO/ZnS composites may be related to the existence of N doping, ZnS/ZnO heterostructure, and covered abundant carbon species on the photocatalyst surface, which causing high absorption efficiency of light, efficient separation of electron-hole pairs, and quick surface reaction in doped ZnO.

INTRODUCTION

Semiconductor photocatalysis is an efficient approach for environmental decontamination by the chemical utilization of solar energy [1–4], which is capable of converting the toxic and nonbiodegradable organic compounds into carbon dioxide and inorganic constituents. Among the various semiconductors applied, TiO_2 is the most frequently employed photocatalyst owing to its cheapness, nontoxicity, and structural stability [5–8]. Currently, researchers show that ZnO has better activity in photocatalytic degradation of some organic contaminants compared to that of TiO_2 [9–11]. However, ZnO semiconductor has a wide band gap of about 3.37 eV, which results in that it is effective only under irradiation of UV-light region and suffers from low efficiency under visible light illumination. Thus, only approximately 3%–5% UV light of the solar energy that reaches the earth can be utilized for photocatalytic reactions when ZnO is used as the catalyst. Furthermore, the fast recombination of photogenerated electron-hole pairs also needs to be solved for its application [12]. Therefore, improving photocatalytic performance of ZnO by modification to reduce the band gap to make absorption in the visible region possible (use sunlight more efficiently) and to inhibit recombination of photogenerated electron-hole pairs has become a hot topic among researchers in recent years [13, 14].

Numerous efforts have been developed to overcome the drawbacks, such as depositing metals on ZnO surface [15, 16], doping with metals (Co^{2+}, Mg^{2+}) or nonmetal ions (C, S, N) [14,17–19], or combining ZnO with another semiconductor [20, 21]. Recently, nonmetal ions doped ZnO photocatalysts have attracted much attention in the photocatalytic processes owing to that it can improve photocatalytic activity by enhancing absorption of light and transport of photogenerated charge carriers [19, 22]. However, to the best of our knowledge, previous researches mostly regarded onefold modification of ZnO with non metal ions such as C, N, S co-doped ZnO or C, N codoped ZnS photocatalysts [14, 23]; few studies have been done on comodification with doping of nonmetal ions and coupling with another semiconductor to enhance photocatalytic performance of semiconductor. Therefore, in the present work, we synthesized the N-doped ZnO/ZnS photocatalysts with visible-light response via a simple heat-treatment approach using L-cysteine as N and S source. Anthraquinone dye (reactive brilliant blue KN-R) is employed as the model contaminants to evaluate the photocatalytic activity of as-synthesized samples under sunlight illumination. The N-doped ZnO/ZnS synthesized by this method shows better photocatalytic activity as compared to that of pure ZnO.

EXPERIMENTAL DETAILS

Synthesis

2.875 g $ZnSO_4 \cdot 7H_2O$ (0.01 mol), 0.8 g NaOH (0.02 mol), and desired amount of L-cysteine (the molar ratio of Zn to L-cysteine is 10 : 1, 5 : 1, 1 : 1, and 1 : 5) were dissolved in 40 mL deionized water. After 10 min of stirring, the reaction mixture was evaporated on water bath at 90°C until dry mixture was obtained, followed by transferring into an annealing furnace and maintained at 300, 400, 500, and 600°C for 2 h in nitrogen atmosphere. After calcination, the powder was collected, washed with deionized water and anhydrous alcohol for three times, and dried at 80°C for 6 h to obtain N-doped ZnO/ZnS composites. For comparison, pure ZnO was prepared by the same process with synthesis of N, S co-doped ZnO composite in absence of L-cysteine and calcined at 400°C 2 h in this process.

Characterizations

The products were characterized by X-ray powder diffraction on a Shimadzu XRD-6100 X-ray diffractometer with a graphite monochromatized CuKa radiation ((λ = 1.5418 Å).). UV-visible diffuse reflectance spectra were recorded with a Varian Cary-100 spectrophotometer and barium sulfate was used as a standard. The surface structure of the as-prepared sample was determined by X-ray photoelectron spectroscopy (XPS) and was performed using VG EscaLab 250 SYSTEM (Thermo VG) with Al K α radiation (1486.6 eV). The C1s photoelectron peak (binding energy at 284.6 eV) was used as energy reference. The PL spectra of ZnO microcrystal photocatalyst were measured by using a fluorescence spectrophotometer (PE-LS55, USA) equipped with a Xenon lamp at an excitation wavelength of 325 nm.

Photocatalytic Activity Test

The photocatalytic activities of the samples were evaluated by the degradation of anthraquinone dye (reactive brilliant blue KN-R) in an aqueous solution. 200 mL anthraquinone dye aqueous solution with concentration of 20 mg/L was mixed with 20 mg/L catalysts, which was exposed to illumination of 500 W Xe lamp (as simulated sunlight source) with a maximum emission at about 470 nm. Before turning on the lamp, the suspension containing reactive brilliant blue KN-R and photocatalyst were magnetically stirred in a dark condition for 60 min till an adsorption-desorption equilibrium was established. Samples were then taken out regularly from the reactor and centrifuged immediately for separation of any suspended solid. The absorbance A of transparent solution was measured by a 721B spectrophotometer and the A value was used to estimate the photocatalytic degradation rate D of reactive brilliant blue K-NR according to the following equation:

$$D = \frac{(A_0 - A_t)}{A_0} \times 100\%,$$

(1)

where A_0 is the initial absorbance of reactive brilliant blue K-NR, t is the reaction time, and A_t is the absorbance at time t.

RESULTS AND DISCUSSION

Figure 1(a) shows the XRD patterns of the N-doped ZnO/ZnS samples calcined at various temperatures. The main diffraction peaks can be indexed for hexagonal wurtzite ZnO (JCPDS card no. 36-1451) and sphalerite cubic ZnS phase (JCPDS card no. 5-0566) in XRD pattern of powders obtained by calcination at 300–500°C. When the calcination temperature of as-synthesized sample reached 600°C, the characteristic peaks of wurtzite ZnS phase is found, while all the characteristic peaks of sphalerite cubic ZnS disappeared. The XRD pattern of as-synthesized samples with different Zn/N molar ratios is shown in Figure 1(b). It can be seen from Figure 1(b) that intensity of diffraction peaks corresponding to ZnO increases with increasing of Zn/N atomic ratio, while intensity of diffraction peaks corresponding to ZnS decreases. The XRD patterns of the (100), (002), and (101) planes of the samples were shown in Figure 1(c). The peaks of these planes in N-doped ZnO/ZnS samples shift slightly to lower Bragg angle (by 0.07°) as compared to those of pure ZnO. This shift suggests that the oxygen or Zn atoms in the lattice of doped ZnO samples may be substituted by other atoms. We consider that the position of diffraction peaks in doped samples shifts to lower Bragg angle, which may be contributed to the fact that atomic radius of N is greater than O and smaller than Zn; it is suggested that N is substituted on O sites [24].

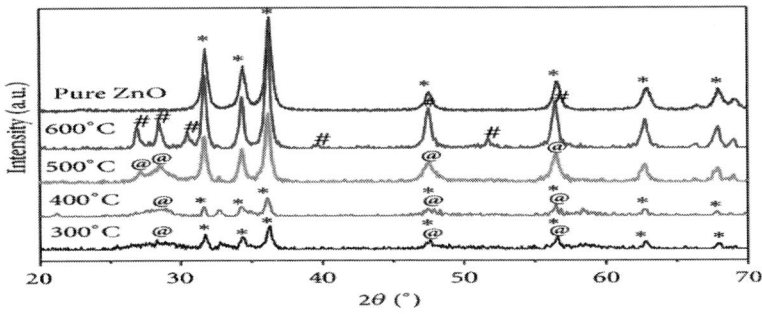

\# ZnS (JCPDS card no. 36-1450)
@ ZnS (JCPDS card no. 5-0566)
* ZnO (JCPDS card no. 36-1451)

(a)

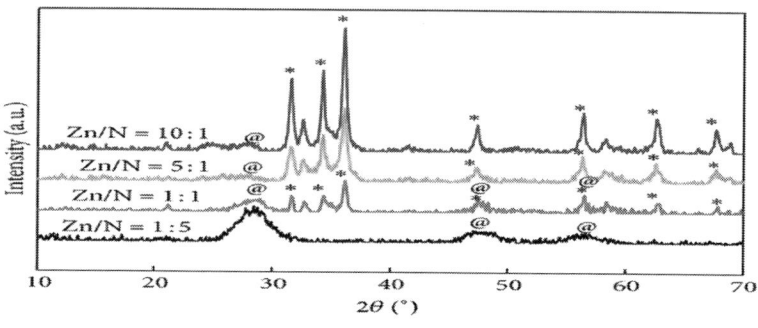

@ ZnS (JCPDS card no. 5-0566)
* ZnO (JCPDS card no. 36-1451)

(b)

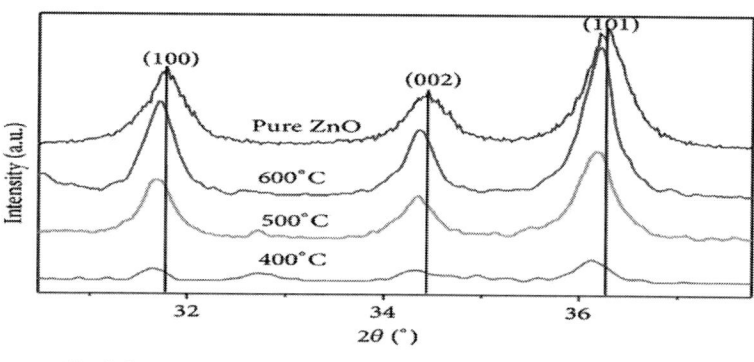

* ZnO (JCPDS card no. 36-1451)

(c)

Figure 1: XRD patterns of as-synthesized samples: (a) samples with Zn/N = 1 : 1 molar ratios calcined at different temperatures, (b) samples with different Zn/N molar ratios calcined at 400°C, and (c) comparison for peaks position between N-doped ZnO/ZnS composites and pure ZnO.

The optical properties of the as-synthesized samples were probed by UV-visible diffuse reflectance spectroscopy. Figure 2 shows the transformed UV-vis absorption spectra of the as-synthesized samples together with the band-gap values, evaluated by linear extrapolation (the intercept on the x -axis gives the value of the band gap). The spectrum of pure ZnO is also included for comparison. From Figure 2, it is clear that N-doped ZnO/ZnS samples show obvious red shift as compared to those of pure ZnO. It is found that the band-gap of N-doped ZnO/ZnS samples decreases with the increasing of the Zn/N molar ratios. Oppositely, the change of band gap with heat treating is irregular, which suggests that the effect of heat treating on structure of photocatalysts may be more complex. Nevertheless, a minimum value of band gap (2.98 eV) can be observed in N-doped ZnO/ZnS sample with Zn/N = 1 prepared by calcination at 400°C. The red shift of the band-gap absorption edge for N-doped ZnO/ZnS can be associated with the formed impurity states in the band gap by the partial substitution of O with other atoms in the crystal lattice of ZnO [22]. Thus, the N-doped ZnO/ZnS can be used as an efficient photocatalyst under sunlight irradiation.

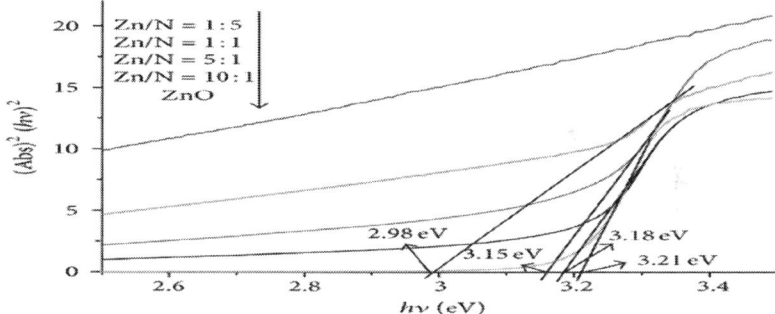

(a)

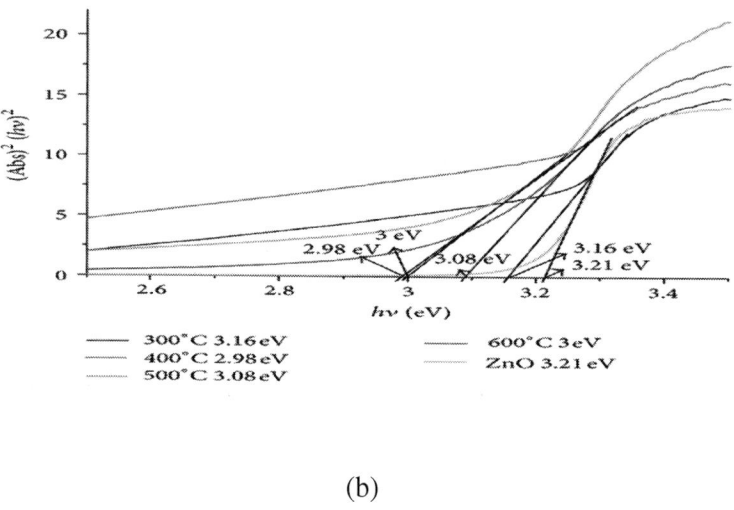

(b)

Figure 2: UV-vis absorption spectra in the band-gap region of samples: (a) samples with different Zn/N molar ratios calcined at; 400°C; (b) samples with Zn/N = 1 : 1 molar ratios calcined at different temperatures.

The PL spectrum of the samples was measured by an excitation wavelength of 325 nm at room temperature and shown in Figure 3. The PL peak at about 390–400 nm was observed for doped ZnO and pure ZnO. The PL peak can be contributed to the recombination of photo-generated electrons and holes [25, 26]. It can be seen that from Figure 3 the N-doped ZnO/ZnS composites showed lower intensity of PL peak located at 390–400 nm than that of pure ZnO, especially the composite with 1 : 1 of Zn/N atomic ratio showed lowest PL intensity. The PL spectra indicate that the effect of doped impurities on the recombination of photo-generated charges depended on the content of impurities. The excessive impurities would form recombination center of photo-generated charges, which improved their recombination. The PL results demonstrated that impurity doping can inhibit the recombination between photogenerated holes and electrons, which is beneficial for the photocatalytic reaction.

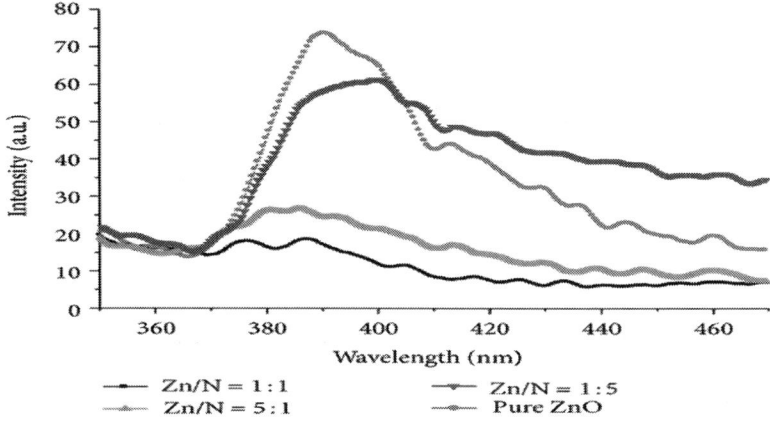

Figure 3: Photoluminescence spectra of samples with different Zn/N molar ratios calcined at 400°C.

The surface composition and chemical state of as-synthesized samples were determined by XPS analysis. Figure 4(a) shows the whole scanning spectrum of as-synthesized samples. The stronger signals of S2p, C1s and a weak signal corresponding to N1s were observed in doped ZnO sample as compared to those of pure ZnO, which indicates S, N, and C elements exist in doped ZnO sample. The C1s XPS spectrum of doped ZnO was fitted to three peaks at 284.5, 286.5 eV, and 288.3 eV (see Figure 4(b)). The peak with a binding energy of 284.5 eV can be assigned to adventitious carbon adsorbed on the surface of the sample [27]. The other two peaks at 286.5 and 288.3 eV can be assigned to the existence of Zn–O–C and C=O of carbonate species, respectively [28, 29]. The peak around 282 eV resulting from carbon interacting with Zn through Zn–C bond formation was not observed [30], which suggests that new chemical state of carbon species (Zn–C) is not formed during the doping process. Even so, the existence of abundant carbon species on the surface of composite still is beneficial for photocatalytic process because it can set in contact with external pollutant molecules.

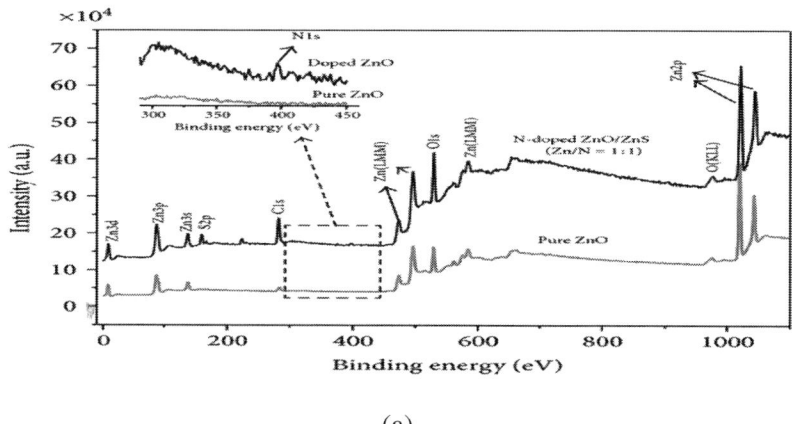

(a)

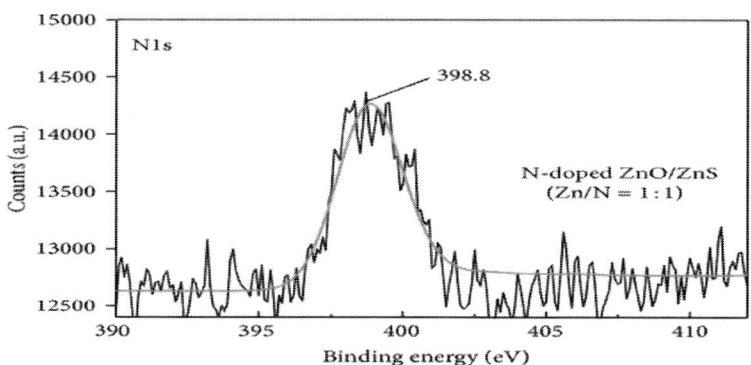

(b)

(c)

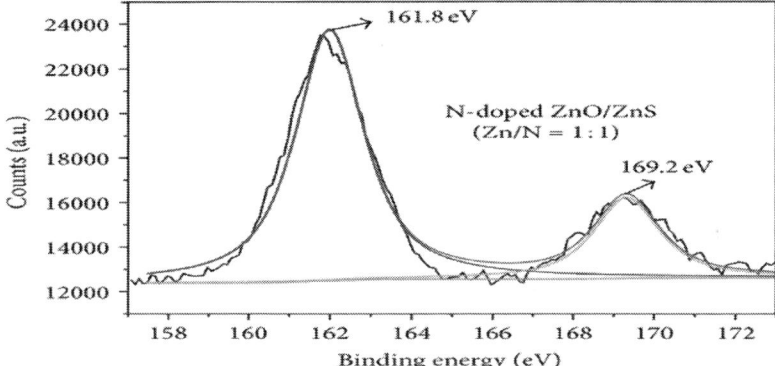

(d)

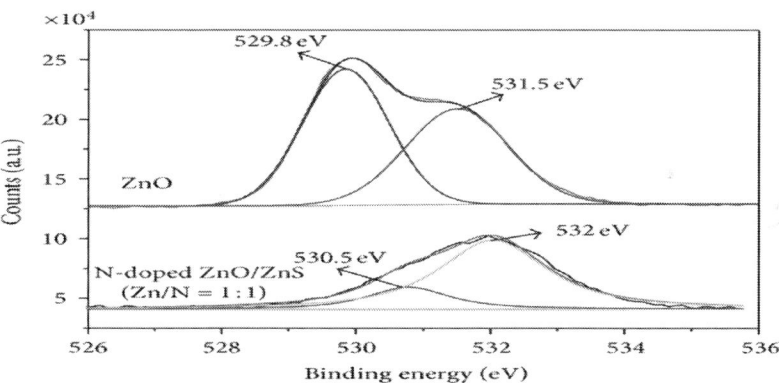

(e)

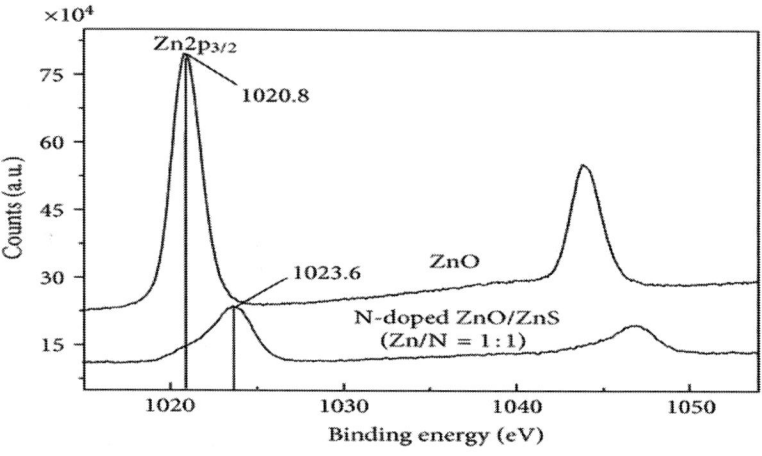

(f)

Figure 4: XPS spectra of samples: (a) XPS survey spectra, (b) C1s, (c) N1s, (d) S2p, (e) O1s, and (f) Zn2p state.

Figure 4(c) shows the high resolution XPS spectra of N1s region for doped ZnO sample and its fitting curves. It can be seen that a peak centered at 398.8 eV for the doped sample is obtained, which can be ascribed to the anionic N^- in the form of N–Zn bonds [31–34]. Thus the XPS data supports the incorporation of N into the ZnO. The substitutional N may be related to the active sites for the photocatalysts [35]. The high resolution XPS spectrum of the S2p region of doped ZnO sample is shown in Figure 4(d). It can be seen that the peak of S2p contains two isolated bands centered at 169.2 and 161.8 eV, which can be attributed to the S^{4+}/S^{6+} and S^{2-} according to the data reported in the literature [28, 36–39]. Thus, based on mentioned XRD, UV-vis, and XPS analysis it can be reasonable to deduce that the N-doped ZnO/ZnS composite was successfully synthesized by simple heat-treated process.

The high resolution XPS spectra of O1s band of as-synthesized samples are shown in Figure4(e). The O1s peak was fitted to two components centered at about 529.8–530.5 eV and 531.5–532.0 eV, respectively. The low binding energy component at 529.8–530.5 eV was attributed to O^{2-} ions of ZnO [40], while another at 531.5–532.0 eV usually corresponds to oxygen in adsorbed O_2 or OH^- groups on the ZnO surface [41–43]. In this study, it is considered that some S, C, and N atoms may bound to oxygen in the sample; the peak at 531.5–532.0 eV may also be assigned to O bound to S, N, or C atom [14]. In addition, from the high resolution XPS spectra of $Zn2p_{3/2}$ (Figure 4(f)), it can be seen that the $Zn2p_{3/2}$ peaks of pure ZnO and doped ZnO appears at 1020.8 and 1023.6 eV, respectively. It is clear that the binding energy of $Zn2p_{3/2}$ peak of doped ZnO is higher as compared to that of the pure ZnO. The shift of $Zn2p_{3/2}$ peak for doped ZnO can be ascribed to the doping or incorporation of N ions into ZnO powders and the existence of Zn–S bond structure [44, 45].

The photocatalytic evaluation of as-synthesized samples is carried out for degradation of reactive brilliant blue KN-R. The straight lines for all reactions were obtained when $\ln(C_0/C_t)$ was plotted against time (see Figure 5), which indicated that the photodegradation process corresponds well to pseudofirst-order kinetics. It can be seen that the N-doped ZnO/ZnS composites exhibited higher photodegradation rate than the pure ZnO. The improvement of photocatalytic activity for N-doped ZnO/ZnS should be related to the nitrogen doping, ZnS/ZnO hetero-structure, and covered carbon species on the photocatalyst surface, which causing high absorption efficiency of light, efficient separation of electron-hole pairs, and quick surface reaction in doped ZnO. The N-doped ZnO/ZnS composite obtained under optimum conditions (Zn/N = 1 : 1 and heat treated at 400°C for 2 h) has a highest degradation rate, and the apparent rate constant k was 0.0053 min^{-1}. The photodegradation rate of the N-doped ZnO/ZnS composite prepared at optimum conditions is about 9.1 times more than that of pure ZnO for reactive brilliant blue KN-R degradation under sunlight irradiation. It is well known that the stability of the ZnO is an important concern for the repeated use of the photocatalysts, so the photodegradative cycling experiments were carried out under sunlight irradiation, and each run lasted 120 min (see Figure 6). The results in Figure 6 show that the photocatalytic activity of N-doped ZnO/ZnS composite does not exhibit any great loss in activity even after five times, which implied that the stability of N-doped ZnO/ZnS composite is suitable for the photodegradation process of pollutants.

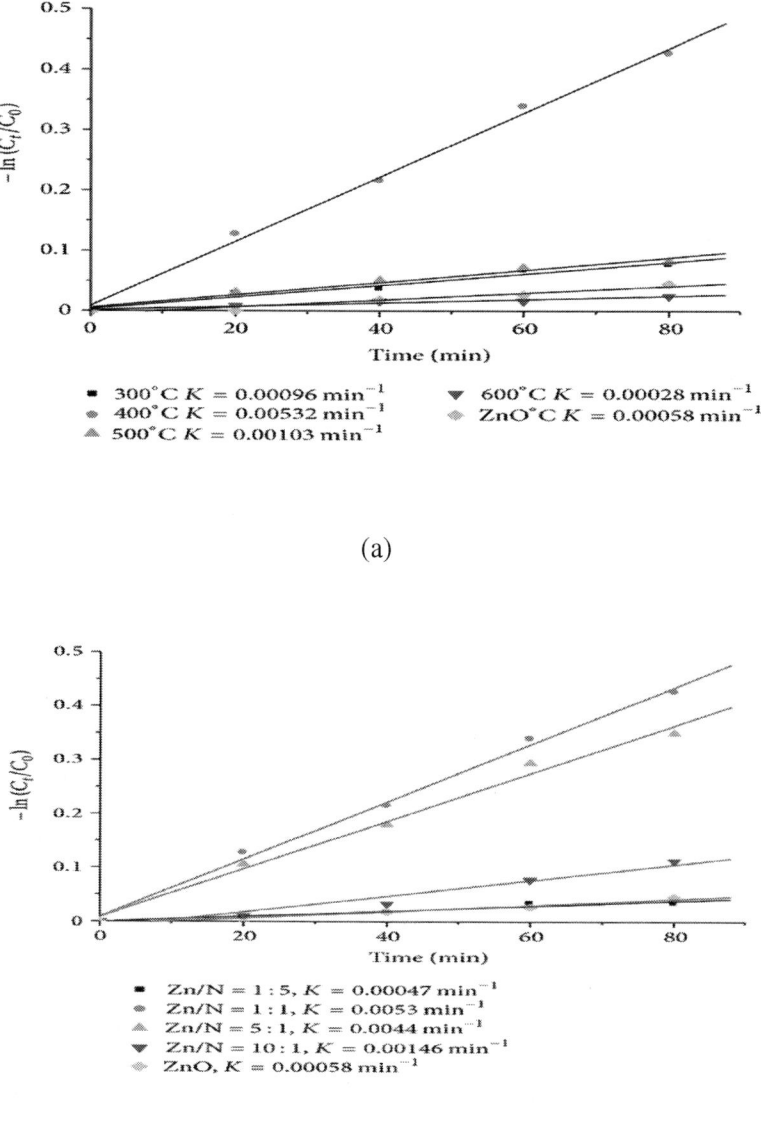

Figure 5: Plots of degradation of KN-R over as-synthesized samples under visible light irradiation; (a) N-doped ZnO/ZnS photocatalysts with Zn/N = 1 prepared at different temperatures for 2 h. (b) N-doped ZnO/ZnS photocatalysts with various Zn/N ratios prepared at 400°C for 2 h.

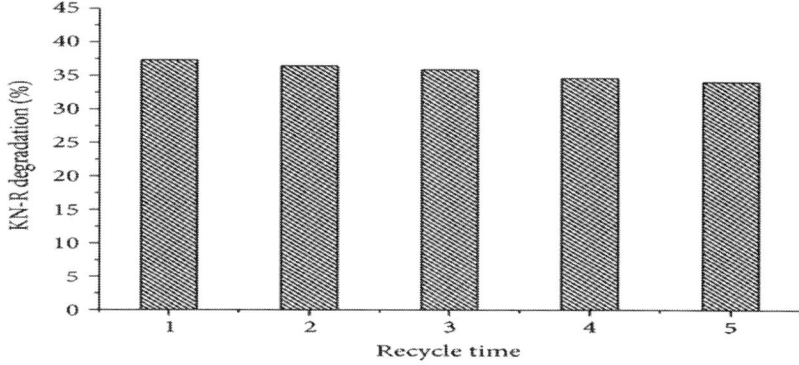

Figure 6: The photocatalytic cycling tests of N-doped ZnO/ZnS composite obtained under optimum conditions (Zn/N =1:1 and heat treated at 400°C for 2 h).

CONCLUSIONS

In this work, the N-doped ZnO/ZnS composites were synthesized by simple heat-treating method using L-cysteine as N and S source. The XRD, XPS, UV-vis DRS, and PL studies showed that the N is incorporated to ZnO/ZnS composites, which shifted the band-gap absorption edge to visible light region and inhibited recombination of photogenerated electron-hole pairs. Hence, the as-synthesized N-doped ZnO/ZnS composites show better photodegradation rate of reactive brilliant blue KN-R as compared to that of pure ZnO under sunlight irradiation. The photo-degradation rate of N-doped ZnO/ZnS composite prepared under optimum conditions (Zn-N = 1:1 and heat treated at 400°C for 2 h) was found to be 9.1-times greater than that of pure ZnO.

ACKNOWLEDGMENTS

This work was supported by the National Natural Science Foundation of China (21243005), Science and Technology Foundation of Liaoning of China (201102111), Program for Key Science and Technology Platform in Universities of Liaoning Province, Program for Liaoning Excellent Talents in University (LGQ-2011-054, LR2011011), and

Science & Technology Public-Benefit Foundation of Liaoning of China ([2011]49).

REFERENCES

1. N. Sobana and M. Swaminathan, "The effect of operational parameters on the photocatalytic degradation of acid red 18 by ZnO," Separation and Purification Technology, vol. 56, no. 1, pp. 101–107, 2007.

2. M. Mrowetz and E. Selli, "Photocatalytic degradation of formic and benzoic acids and hydrogen peroxide evolution in TiO_2 and ZnO water suspensions," Journal of Photochemistry and Photobiology A, vol. 180, no. 1-2, pp. 15–22, 2006.

3. K. Chiang, T. M. Lim, L. Tsen, and C. C. Lee, "Photocatalytic degradation and mineralization of bisphenol A by TiO_2 and platinized TiO_2," Applied Catalysis A, vol. 261, no. 2, pp. 225–237, 2004.

4. H. Lachheb, E. Puzenat, A. Houas et al., "Photocatalytic degradation of various types of dyes (Alizarin S, Crocein Orange G, Methyl Red, Congo Red, Methylene Blue) in water by UV-irradiated titania," Applied Catalysis B, vol. 39, no. 1, pp. 75–90, 2002.

5. A. L. Linsebigler, G. Lu, and J. T. Yates, "Photocatalysis on TiO_2 surfaces: principles, mechanisms, and selected results," Chemical Reviews, vol. 95, no. 3, pp. 735–758, 1995.

6. A. Fujishima, T. N. Rao, and D. A. Tryk, "Titanium dioxide photocatalysis," Journal of Photochemistry and Photobiology C, vol. 1, no. 1, pp. 1–21, 2000.

7. S. Klosek and D. Raftery, "Visible light driven V-doped TiO_2 photocatalyst and its photooxidation of ethanol," Journal of Physical Chemistry B, vol. 105, no. 14, pp. 2815–2819, 2002.

8. C. Kormann, D. W. Bahnemann, and M. R. Hoffmann, "Environmental photochemistry: is iron oxide (hematite) an active photocatalyst? A comparative study: -Fe_2O_3, ZnO, TiO_2," Journal of Photochemistry and Photobiology A, vol. 48, no. 1, pp. 161–169, 1989.

9.   F. Peng, H. Wang, H. Yu, and S. Chen, "Preparation of aluminum foil-supported nano-sized ZnO thin films and its photocatalytic degradation to phenol under visible light irradiation," Materials Research Bulletin, vol. 41, no. 11, pp. 2123–2129, 2006

10. D. Li and H. Haneda, "Morphologies of zinc oxide particles and their effects on photocatalysis," Chemosphere, vol. 51, no. 2, pp. 129–137, 2003.

11. C. Hariharan, "Photocatalytic degradation of organic contaminants in water by ZnO nanoparticles," Applied Catalysis A, vol. 304, pp. 55–61, 2006.

12. M. Romero, J. Blanco, B. Sánchez et al., "Solar photocatalytic degradation of water and air pollutants: challenges and perspectives," Solar Energy, vol. 66, no. 2, pp. 169–182, 1999.

13. Y. Zheng, L. Zheng, Y. Zhan, X. Lin, Q. Zheng, and K. Wei, "Ag/ZnO heterostructure nanocrystals: synthesis, characterization, and photocatalysis," Inorganic Chemistry, vol. 46, no. 17, pp. 6980–6986, 2007.

14. L.-C. Chen, Y.-J. Tu, Y.-S. Wang, R.-S. Kan, and C.-M. Huang, "Characterization and photoreactivity of N-, S-, and C-doped ZnO under UV and visible light illumination,"Journal of Photochemistry and Photobiology A, vol. 199, no. 2-3, pp. 170–178, 2008.

15. J. Liqiang, W. Baiqi, X. Baifu et al., "Investigations on the surface modification of ZnO nanoparticle photocatalyst by depositing Pd," Journal of Solid State Chemistry, vol. 177, no. 11, pp. 4221–4227, 2004.

16. W. Lu, S. Gao, and J. Wang, "One-pot synthesis of Ag/ZnO self-assembled 3D hollow microspheres with enhanced photocatalytic performance," Journal of Physical Chemistry C, vol. 112, no. 43, pp. 16792–16800, 2008.

17. X. Qiu, G. Li, X. Sun, L. Li, and X. Fu, "Doping effects of Co^{2+} ions on ZnO nanorods and their photocatalytic properties," Nanotechnology, vol. 19, no. 21, Article ID 215703, 8 pages, 2008.

18. X. Qiu, L. Li, J. Zheng, J. Liu, X. Sun, and G. Li, "Origin of the enhanced photocatalytic activities of semiconductors: a case study of ZnO doped with Mg^{2+}," Journal of Physical Chemistry C, vol. 112, no. 32, pp. 12242–12248, 2008.

19. J. Lu, Q. Zhang, J. Wang, F. Saito, and M. Uchida, "Synthesis of N-Doped ZnO by grinding and subsequent heating ZnO-urea mixture," Powder Technology, vol. 162, no. 1, pp. 33–37, 2006.

20. M. L. Zhang, T. C. An, X. H. Hu, C. Wang, G. Y. Sheng, and J. M. Fu, "Preparation and photocatalytic properties of a nanometer $ZnO-SnO_2$ coupled Oxide," Applied Catalysis A, vol. 260, no. 2, pp. 215–222, 2004.

21. R. S. Mane, W. J. Lee, H. M. Pathan, and S. H. Han, "Nanocrystalline TiO_2/ZnO thin films: fabrication and application to dye-sensitized solar cells," Journal of Physical Chemistry B, vol. 109, no. 51, pp. 24254–24259, 2005.

22. D. Li and H. Haneda, "Synthesis of nitrogen-containing ZnO powders by spray pyrolysis and their visible-light photocatalysis in gas-phase acetaldehyde decomposition," Journal of Photochemistry and Photobiology A, vol. 155, no. 1-3, pp. 171–178, 2003.

23. M. Muruganandham and Y. Kusumoto, "Synthesis of N, C codoped hierarchical porous microsphere ZnS as a visible light-responsive photocatalyst," Journal of Physical Chemistry C, vol. 113, no. 36, pp. 16144–16150, 2009.

24. S. S. Shinde, C. H. Bhosale, and K. Y. Rajpure, "Photocatalytic degradation of toluene using sprayed N-doped ZnO thin films in aqueous suspension," Journal of Photochemistry and Photobiology B, vol. 113, pp. 70–77, 2012.

25. C. Chandrinou, N. Boukos, C. Stogios, and A. Travlos, "PL study of oxygen defect formation in ZnO nanorods," Microelectronics Journal, vol. 40, no. 2, pp. 296–298, 2009.

26. G. R. Li, C. R. Dawa, Q. Bu et al., "Electrochemical self-assembly of ZnO nanoporous structures," Journal of Physical Chemistry C, vol. 111, no. 5, pp. 1919–1923, 2007.

27. C. S. Gopinath, S. G. Hegde, A. V. Ramaswamy, and S. Mahapatra, "Photoemission studies of polymorphic $CaCO_3$ materials," Materials Research Bulletin, vol. 37, no. 7, pp. 1323–1332, 2002.

28. H. Q. Sun, Y. Bai, Y. P. Cheng, W. Q. Jin, and N. P. Xu, "Preparation and characterization of visible-light-driven carbon-sulfur-codoped TiO_2 photocatalysts," Industrial & Engineering Chemistry Research, vol. 45, no. 14, pp. 4971–4976, 2006.

29. E. Papirer, R. Lacroix, J. B. Donnet, G. Nansé, and P. Fioux, "XPS study of the halogenation of carbon black—part 2. Chlorination," Carbon, vol. 33, no. 1, pp. 63–72, 1995.

30. S. Cho, J. W. Jang, J. S. Lee, and K. H. Lee, "Carbon-doped ZnO nanostructures synthesized using vitamin C for visible light photocatalysis," CrystEngComm, vol. 12, no. 11, pp. 3929–3935, 2010.

31. J. G. Ma, Y. C. Liu, R. Mu et al., "Method of control of nitrogen content in ZnO films: structural and photoluminescence properties," Journal of Vacuum Science & Technology B, vol. 22, no. 1, pp. 94–98, 2004

32. X. Wang, J. C. Yu, Y. Chen, L. Wu, and X. Fu, "ZrO_2-modified mesoporous nanocrystalline TiO2−xNx as efficient visible light photocatalysts," Environmental Science and Technology, vol. 40, no. 7, pp. 2369–2374, 2006.

33. Y. F. Mei, R. K. Y. Fu, G. G. Siu et al., "Nitrogen binding behavior in ZnO films with time-resolved cathodoluminescence," Applied Surface Science, vol. 252, no. 23, pp. 8131–8134, 2006.

34. S. Sato, R. Nakamura, and S. Abe, "Visible-light sensitization of TiO_2 photocatalysts by wet-method N doping," Applied Catalysis A, vol. 284, no. 1-2, pp. 131–137, 2005.

35. R. Asahi, T. Morikawa, T. Ohwaki, K. Aoki, and Y. Taga, "Visible-light photocatalysis in nitrogen-doped titanium oxides," Science, vol. 293, no. 5528, pp. 269–271, 2001.

36. J. C. Yu, W. Ho, J. Yu, H. Yip, K. W. Po, and J. Zhao, "Efficient visible-light-induced photocatalytic disinfection on sulfur-doped nanocrystalline titania," Environmental Science and Technology, vol. 39, no. 4, pp. 1175–1179, 2005.

37. T. Ohno, M. Akiyoshi, T. Umebayashi, K. Asai, T. Mitsui, and M. Matsumura, "Preparation of S-doped TiO_2 photocatalysts and their photocatalytic activities under visible light," Applied Catalysis A, vol. 265, no. 1, pp. 115–121, 2004.

38. T. Ohno, T. Tsubota, Y. Nakamura, and K. Sayama, "Preparation of S, C cation-codoped $SrTiO_3$ and its photocatalytic activity under visible light," Applied Catalysis A, vol. 288, no. 1-2, pp. 74–79, 2005.

39.   A. Zaleska, P. Górska, J. W. Sobczak, and J. Hupka, "Thioacetamide and thiourea impact on visible light activity of TiO_2," Applied Catalysis B, vol. 76, no. 1-2, pp. 1–8, 2007.

40.   S. Anandan, A. Vinu, K. L. P. S. Lovely et al., "Photocatalytic activity of La-doped ZnO for the degradation of monocrotophos in aqueous suspension," Journal of Molecular Catalysis A, vol. 266, no. 1-2, pp. 149–157, 2007.

41.   X. H. Wang, S. Liu, P. Chang, and Y. Tang, "Influence of S incorporation on the luminescence property of ZnO nanowires by electrochemical deposition," Physics Letters A, vol. 372, no. 16, pp. 2900–2903, 2008.

42.   B. Stypula and J. Stoch, "The characterization of passive films on chromium electrodes by XPS," Corrosion Science, vol. 36, no. 12, pp. 2159–2167, 1994.

43.   C. Shifu, Z. Sujuan, L. Wei, and Z. Wei, "Preparation and activity evaluation of p–n junction photocatalyst NiO/TiO_2," Journal of Hazardous Materials, vol. 155, no. 1-2, pp. 320–326, 2008.

44.   C. Shifu, Z. Wei, Z. Sujuan, and L. Wei, "Preparation, characterization and photocatalytic activity of N-containing ZnO powder," Chemical Engineering Journal, vol. 148, no. 2-3, pp. 263–269, 2009.

45.   J. W. Jung, H. C. Lee, and J. S. Wang, "A study on the double insulating layer for HgCdTe MIS structure," Thin Solid Films, vol. 290-291, pp. 18–22, 1996.

Synthesis of Novel ZnO Having Cauliflower Morphology for Photocatalytic Degradation Study

Dipak Nipane[1], S. R. Thakare[1], and N. T. Khati[2]

[1]Nano Technology Lab, Science College, Congress Nagar, Nagpur, Maharashtra 12, India

[2]Priyadarshini College of Engineering and Technology, Nagpur, Maharashtra 19, India

ABSTRACT

ZnO nanowire morphology has been widely studied due to its unique material properties and excellent performance in electronics, optics, and photonic. Recently, photocatalytic applications of ZnO nanowire are creating an increasing interest in the environmental applications.

This paper presents a low-cost and ecofriendly synthesis of ZnO with cauliflower morphology and its effectiveness in photocatalysis.

INTRODUCTION

Organic dyes are very important and widely used in different industries such as textile, rubber, and plastic and hence one of the largest group of pollutants released into wastewaters [1]. They have caused several environmental contaminations, affecting human survival and developments. Degradation and removal of them are a great challenge for protecting the environment. However, the routine techniques for treating organic dyes are usually ineffective and costly to remove pollutions from water [2]. Nanomaterials have attracted high interest due to their noticeable performance in electronics, optics, and photonics. One-dimensional (1D) nanostructures such as nanowire, nanorods, nanofibers, nanobelts, and nanotubes have been of intense interest in both academic research and industrial applications. They also play an important role as interconnectors and functional units in the fabrication of electronics, optoelectronics, electrochemical, and electromechanical nanodevices [3]. For their application in solar energy conversion and environmental purification, among various semiconductor photocatalysts, TiO_2 is much known for its strong oxidizing power and nontoxicity. But the major limitation in semiconductor photocatalysis is the relatively low value of the overall quantum efficiency, because of the high recombination rate of photo-induced electron-hole pairs at or near its surface.

ZnO is a semiconductor material with direct wide band gap energy (3.37eV) and a large exciton binding energy (60meV) at room temperature. ZnO is currently attracting worldwide intense interests because of its importance in fundamental studies and its numerous applications especially as optoelectronic materials [4, 5], UV lasers [6].Light-emitting diodes [7], solar cells [8], nanogenerators [9], gas sensors [10], photodetectors [11], and photocatalyst [12].

Photocatalysis is the promising process for environmental protection because of being able to oxidize low concentrations of organic pollutants into benign products [13–15]. Figure 1 shows mechanism of photocatalytic process. There are number of semiconductors that could be used as photocatalyst, such as TiO_2, ZnO, and WO_3. Although

TiO_2 is the most widely investigated photocatalyst, ZnO has also been considered as suitable alternative of TiO_2 because of its comparability with TiO_2 band gap energy and its relatively lower cost of production. [16–18]. Moreover, ZnO has been reported to be more photoactive than TiO_2 [19–22] due to its higher efficiency of generation and separation of photoinduced electrons and holes [16, 23, 24] to interactive bacteria, viruses, and for the degradation of environmental pollutants.

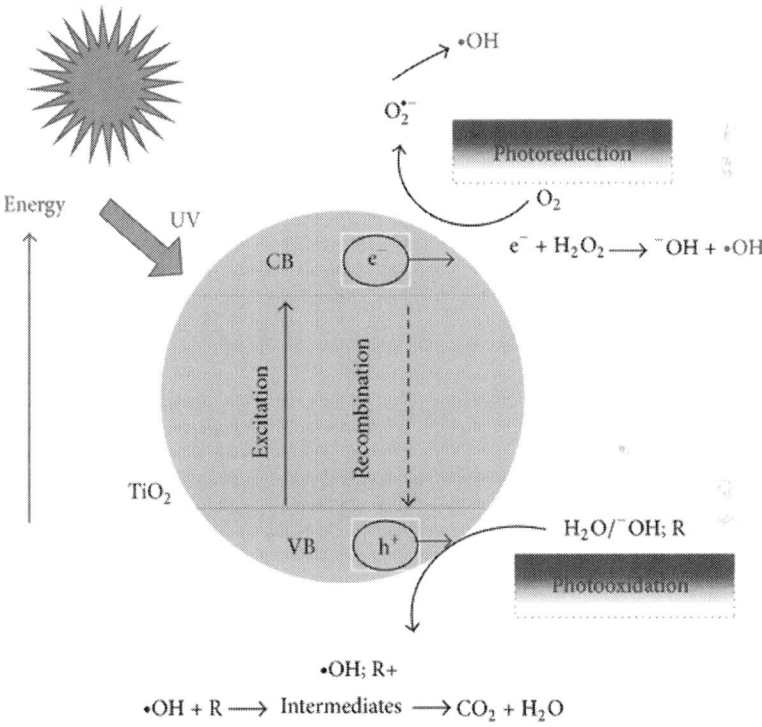

Figure 1: A schematic of the principle of photocatalysis from [25].

In this paper, the synthesis of low-cost and ecofriendly ZnO, an emphasis on cauliflower morphology of ZnO, is used in photocatalysis. We have presented the characterization of ZnO cauliflower and compare its properties with other semiconductors like ZnO nanowire and TiO_2.

EXPERIMENTAL DATA

Synthesis of ZnO Nanoparticle

Sol-gel process has many advantages when compared to vapor phase and hydrothermal processes, because of its low cost, low temperature, scalability, and ease of handling. Generally, sol-gel reaction occurs at relatively low temperature (<200); therefore, to save energy sol-gel elemental reaction was used for the preparation of nanocauliflower ZnO particle. The ZnO crystalline powder was prepared by the following process with the use of zinc acetate as metal precursor and starch as stabilizer. Zinc acetate with purity (99%) and starch having purity (99%) were purchased from Merck India and used without further purification. Ammonia was purchased from S.D. Fine chem. Ltd. with purity (27%). The aqueous solution (0.1mol/L) of zinc acetate dihydrate was prepared with deionized water, and 4% starch was added as stabilizer. The pure ammonia was slowly added into zinc acetate solution at room temperature under stirring rate of 4000rpm, which resulted in the formation of a white suspension. Then, it refluxed for half an hour at 100°C. The suspension was then separated with a centrifuge 5000rpm and washed properly with plenty of distilled water and was washed with absolute alcohol at last to remove the unreacted chemicals.

Photocatalytic Activity of ZnO Nanocauliflower

Photocatalytic activities of nano-ZnO sample were evaluated by photocatalytic degradation of MB under UV lamp (Philips 18W) using as a light source. A beaker of 100mL was used as reaction vessel, and the distance from the lamp to the solution was 12.5cm. The experiments were performed at room temperature as follows: 0.1g of photocatalyst was added into 100mL MB solution of 2ppm (2mg/Liter), and the pH of the initial solution was about 7. Prior to irradiation, the suspensions were magnetically stirred in the dark for 1h to ensure the adsorption/desorption equilibrium between the photocatalyst powders and the solution. At given time intervals of 10 minute, 4mL of suspensions were sampled and centrifuged to remove the photocatalyst powders.

Evaluation of photocatalytic activities of photocatalyst was conducted by recording the variations of absorption band maximum through a UV-Visible spectrophotometer (Shimadzu UV-1800). MB concentration was analyzed by recording the variations of the absorption band maximum (663nm).

RESULT AND DISCUSSION

XRD

The X-ray diffraction analysis of prepared ZnO nanocauliflowers was carried to identify the product (Figure2). All diffraction peaks are indexed with the corresponding planes of ZnO. The XRD spectra showing the highest intense peak at 36.16, 31.73, and 34.40 which are the crystal plane of ZnO. The low intensity peaks at 47.46, 56.47, and 62.83 which match well with the JCPDS File no. 00-036-1451. No other diffraction peaks arising from metallic Zn or $Zn(OH)_2$ are present in the XRD pattern, which indicates the high phase purity of the synthesized sample. The crystal size of products calculated by Scherrer formula was 50 ± 100 ?nm.

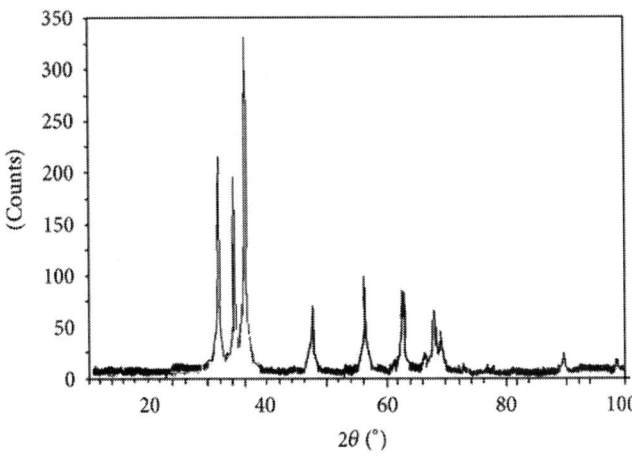

Figure 2: X-ray diffraction of cauliflower ZnO nanocauliflowers.

Morphology

Texture, thickness, and crystal size of ZnO also affect the quality of ZnO [26–29]. Wu et al. [27] studied the effect of seed layer characteristics on the synthesis of ZnO nanowire. The effect of the starch on the morphology of products has been reported in the literature [28]. Figure 3 shows the role of starch in the preparation of ZnO. So, a control experiment without the addition of starch was done here, and the SEM images of the products are exhibited in Figure 4(a) it shows that the morphology of products is rocks and same experiment with addition of starch was done and the SEM images of products are exhibited in Figure 4(b). It shows cauliflower like morphology. This may attribute to the larger surface-to-volume ratio of ZnO cauliflowers than that of rocks, which helps to increase the photocatalytic reactions sites and promote the electron-hole separation. Additionally, the PL results indicate that ZnO nanocauliflowers have few defects compared with ZnO nanorods. Defects may serve as recombination centers for photoexcited electron-hole pairs during photocatalysis; therefore, the decrease of defects implied the decrease of photocatalytic activity. From the TEM (Figure 5) image, it is confirmed that the particle size between 50 and 100nm agrees well with XRD analysis. TEM images confirmed the connectivity between cauliflowers which is observed in SEM images. The SEM and TEM images of ZnO show that the growth of particle is very well organized. Nuclei formation and growth of the particle simultaneously occur, and hence it is appears like cauliflower. Starch is biological molecule having symmetric orientation and may not allow the growth of the particles in random manner. This is may be advantages to use of the biological molecule as a stabilizer of nanoparticle having a specific orientation as well as beautiful morphological nature having tailored properties of specific application [28].

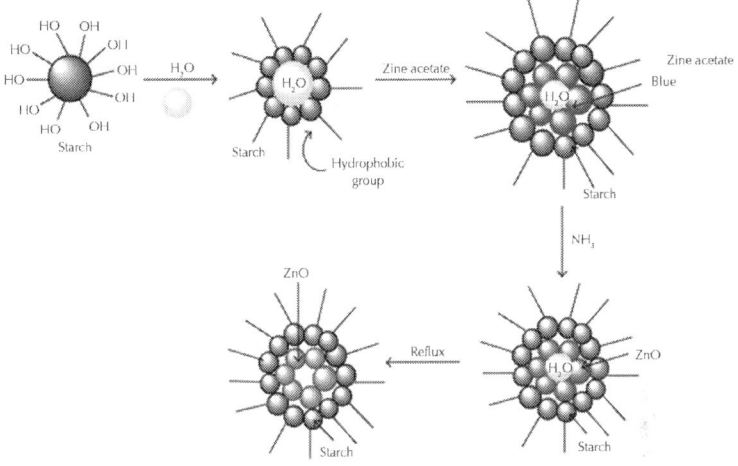

Figure 3: Role of starch in the synthesis of cauliflower ZnO.

(a)

(b)

Figure 4: (a) SEM image of ZnO nanoparticle without starch and (b) SEM image of ZnO nanoparticle with starch are well arranged with specific size.

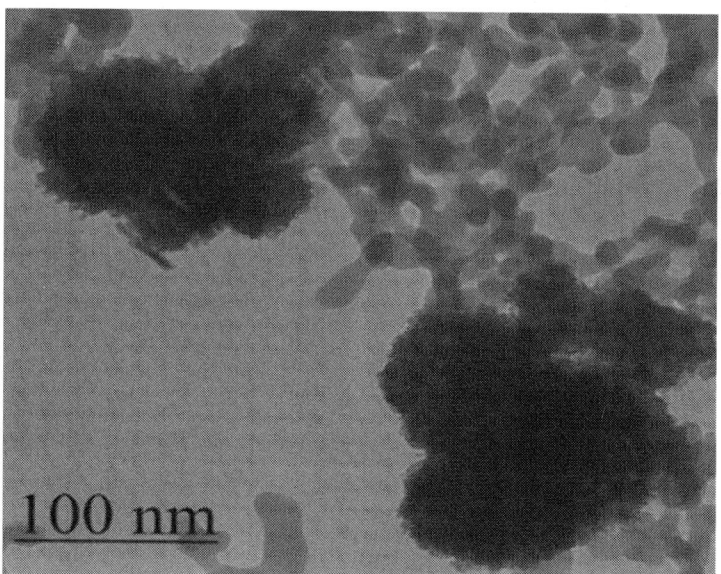

(a)

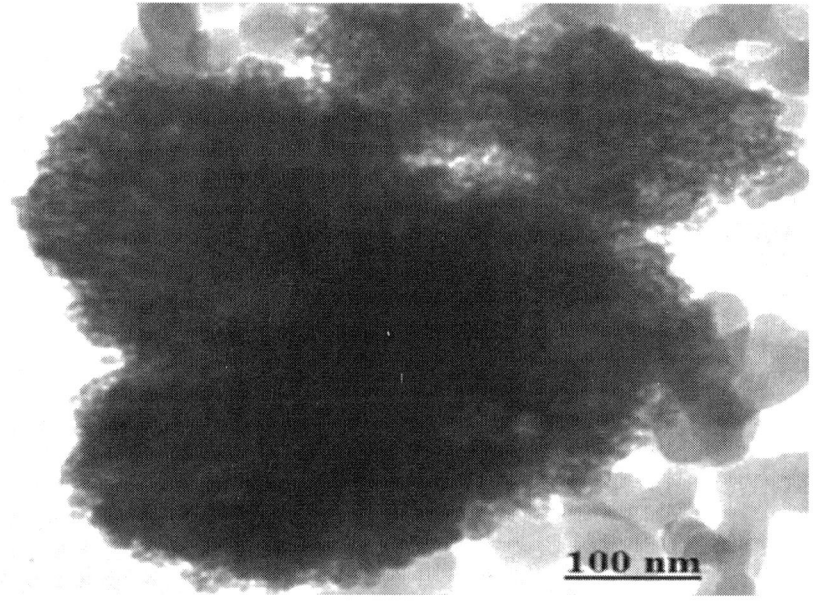

100 nm

(b)

Figure 5: TEM image of ZnO nanocauliflowers with starch.

FTIR

Figure 6 shows the FTIR of the ZnO nanoparticle prepared by the sol-gel method, in the range of 4000–600cm^{-1}. A broad absorption band was observed at around 616, 716, and 2913cm^{-1}; it clearly represented bonding between Zn–O and C–H. There were several small absorption bands at 921, 1024, 1062, and 3316cm^{-1} that are likely related to CO_2 (C–O) and H_2O (O–H) absorbed from the atmosphere (air), and these absorption bands can be neglected. The FTIR results show the high purity of the obtained ZnO nanoparticle.

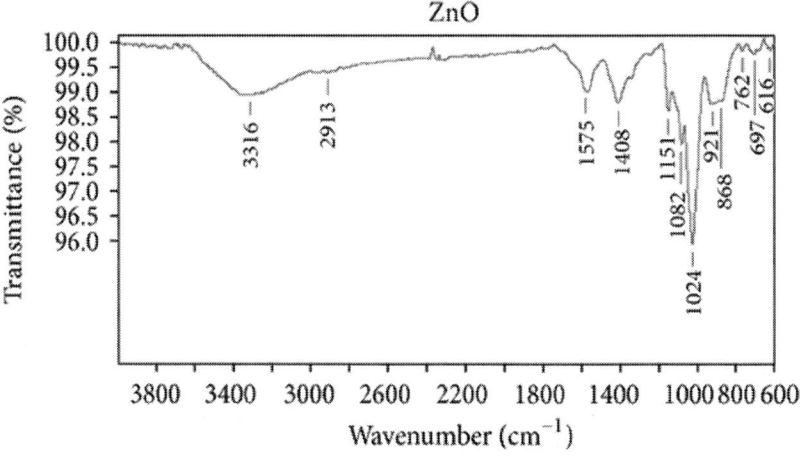

Figure 6: FTIR spectrum of ZnO nanocauliflowers.

Optical Properties

The room temperature UV absorption spectra of ZnO nanoparticle are shown in Figure 7. The spectrum reveals a characteristic absorption peak of ZnO at wavelength of 378? nm which can be assigned to the intrinsic band gap absorption of ZnO due to the electron transitions from the valence band to the conduction band (O_{2p} Zn_{3d}). In addition, this sharp peak shows that the particles are in nanosize, and the particle size distribution is narrow. It is clearly shown that the maximum peak in the absorbance spectrum does not correspond to the true optical band gap of the ZnO nanoparticle. A common way to obtain the band gap from absorbance spectra is to get the first derivative of the absorbance with respect to wavelength and find the maximum in the derivative spectrum at the lower energy sides. It indicates a band gap of 3.26eV. for the nanocauliflower ZnO particle. The good absorption of the ZnO nanoparticle in the UV region proves the applicability of this product in photocatalytic applications.

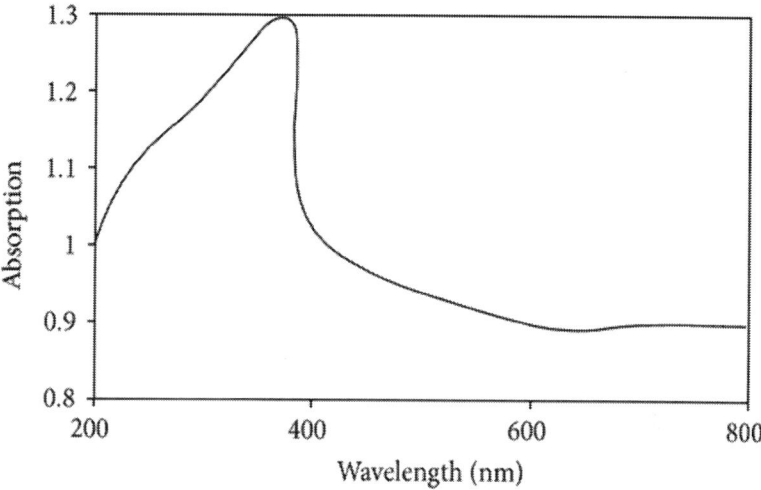

Figure 7: UV-visible absorbance spectrum of ZnO nanocauliflower particle.

PL Spectrum

Optical properties of functional nanomaterials are of importance considering further applications in nanoelectronic devices. Therefore, the optical properties of the prepared ZnO nanocauliflower were further investigated by PL spectroscopy. Figure 8 shows a typical room temperature photoluminescence (PL) spectrum of ZnO nanorods and nanocauliflower with excitation wavelength 325nm at room temperature. The spectrum exhibits two bands including a strong ultraviolet emission at 378nm (or 3.26eV) and a weak spectral band in visible region. The UV emission was contributed to the near band-edge emission of the wide band gap of ZnO. Visible emission is due to the presence of various defects such as oxygen vacancies.

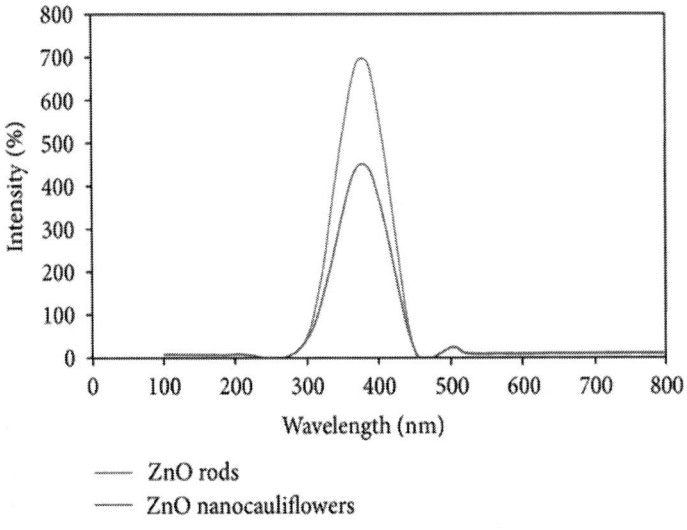

—— ZnO rods
—— ZnO nanocauliflowers

Figure 8: Room temperature photoluminescence spectrum of ZnO nanocauliflower and rods ($\lambda_{exc} = 325$?nm).

PHOTOCATALYTIC DEGRADATION OF METHYLENE BLUE

The various organic pollutants that have been tested and removed under UV light illumination include methylene blue, monocrotophos, and diphenylamine. ZnO nanowires used as photocatalysts have been recently reported by many research groups [30–33]. Sugunan et al. [30] described a continuous flow water purification system by the fabrication of ZnO nanowires. Baruah et al. [34] reported a fast crystallization ZnO nanorods synthesis method to increase the surface defect of the ZnO nanowires, an increase the surface defects and vacancies are capable of exhibiting visible light photocatalysis even without doping with transition metal [34]. In this paper, photocatalytic activity testing was carried out in UV light and was degraded 2ppm of methylene blue solution.

The control analyses show that the degradation of MB is negligible in the absence of ZnO catalysts. The degradations of the MB in the presence of ZnO nanorods (Figure 9(a)) and ZnO nanocauliflowers (Figure9 (b)).were measured by UV irradiation (λ_{max}=365nm) at room temperature at various times. The result indicates that the ZnO cauliflowers show higher photoactivity after 120 minutes (Figure 9(c)). We also tested the photocatalytic activity for ZnO nanocauliflowers, ZnO nanowires, and TiO$_2$. Figure 10 revealed that ZnO nanocauliflowers accelerate the photocatalytic degradation process as compared to ZnO nanowires and TiO$_2$ nanoparticles. It is due to surface properties, such as surface defects and oxygen vacancies of photocatalysts, that play a significant role in the photocatalytic activity and also morphology of ZnO that affects the photocatalytic activity, SEM images for prepared ZnO showed cauliflower-like structure which may absorb more light because of surface modification as compared to nanowires and enhance the photocatalytic activity.

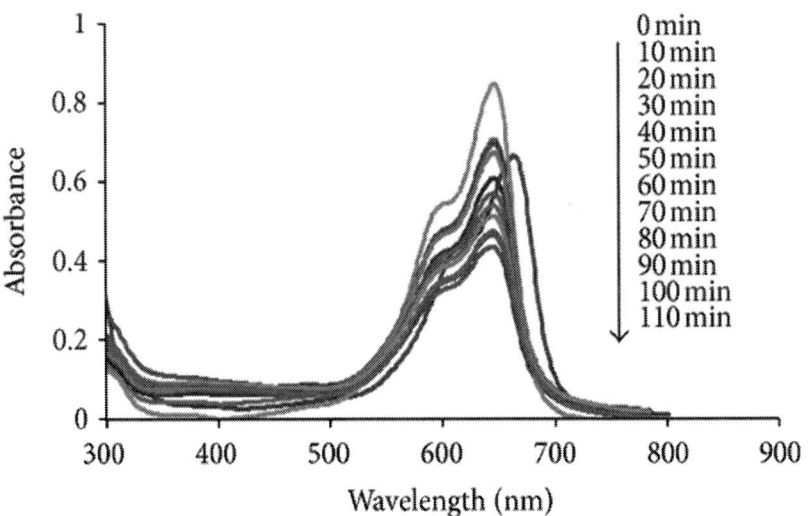

(a)

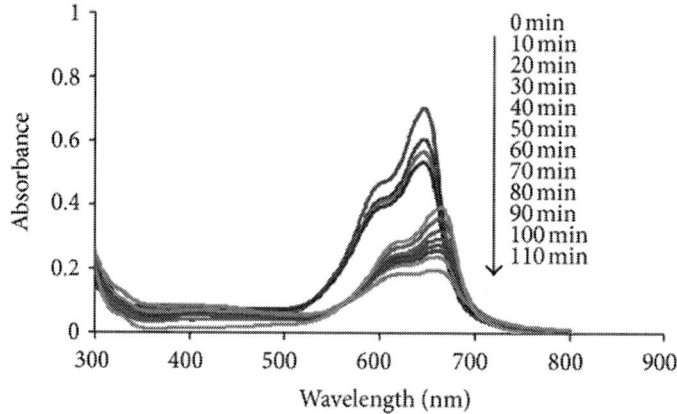

(b)

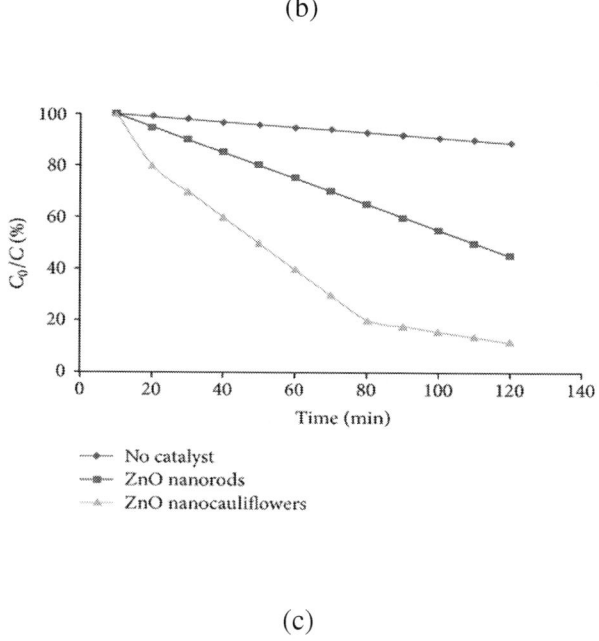

(c)

Figure 9: (a) Photocatalytic degradation of MB at UV radiation by using ZnO nanorods, (b) ZnO nanocauliflowers, and (c) photodegradation rate in presence of catalyst (ZnO nanorods and ZnO nanocauliflowers) (C_0 is the initial concentration of MB, and C is the reaction concentration of MB at time t in minute.).

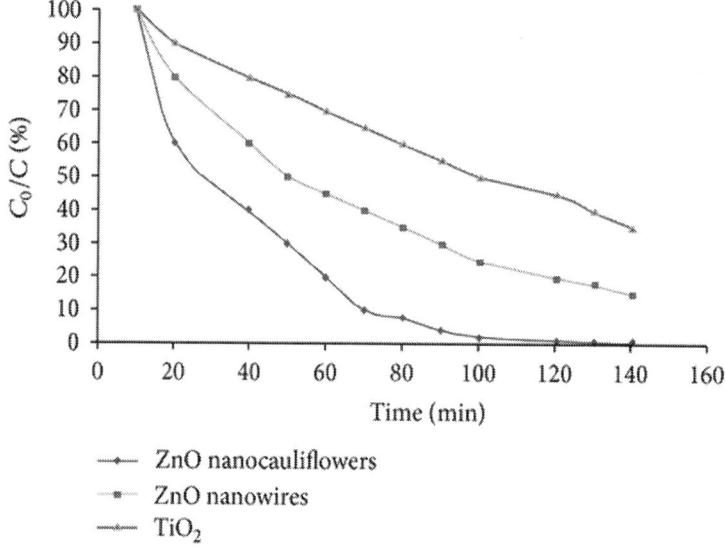

- —◆— ZnO nanocauliflowers
- —■— ZnO nanowires
- —▲— TiO$_2$

Figure 10: Photocatalytic degradation of MB at UV radiation by using ZnO cauliflowers (blue), ZnO nanowires (green), and TiO$_2$ catalyst (C_0 is the initial concentration of MB, and C is the reaction concentration of MB at time t in minute.).

To further investigate the mechanism, we studied the photocatalytic reaction dynamics. It has been well established that heterogeneous photocatalysis by semiconductors follows the Langmuir-Hinshelwood (LH) dynamic model:

$$\frac{1}{R} = \frac{1}{kK} \times \frac{1}{C} + \frac{1}{k},$$

(1)

Where R is the reaction rate, k is the reaction rate constant, K is the adsorption equilibrium constant, and C_0 is the initial dye concentration. When C_0 is small, the reaction follows first-order kinetics with a rate equation in the form

$$-\ln \frac{C}{C_0} = kKt = k't,$$

(2)

Where C is the dye concentration at time t and k' is the apparent first-order rate constant.

CONCLUSIONS

This paper provides an overview of the synthesis, characterizations, and application of ZnO nanocauliflowers. The sol-gel method is simple, low cost, and efficient, and it has received an increase attention. Through changing the morphologies, starch play an important role in the growth of ZnO nanocauliflowers. Due to the unique properties of material, ZnO nanocaulifliowers are attractive for the number of potential application such as photocatalysis, solar cell, sensor, and generators. Among the application of the ZnO nanocauliflowers, photocatalysis is used for environmental protection. Based on this paper, ZnO nanocauliflowers promise to be one of the most important materials in photocatalytic as well as other application.

REFERENCES

1. F. Han, V. S. R. Kambala, M. Srinivasan, D. Rajarathnam, and R. Naidu, "Tailored titanium dioxide photocatalysts for the degradation of organic dyes in wastewater treatment: a review," Applied Catalysis A, vol. 359, no. 1-2, pp. 25–40, 2009.

2. L. Y. Yang, S. Y. Dong, J. H. Sun, J. L. Feng, Q. H. Wu, and S. P. Sun, "Microwave-assisted preparation, characterization and photocatalytic properties of a dumbbell-shaped ZnO photocatalyst," Journal of Hazardous Materials, vol. 179, no. 1–3, pp. 438–443, 2010.

3. G. C. Yi, C. Wang, and W. I. Park, "ZnO nanorods: synthesis,characterization and applications,"Semiconductor Science and Technology, vol. 20, pp. S22–S34, 2005.

4. F. Lu, W. Cai, and Y. Zhang, "ZnO hierarchical micro/ nanoarchitectures: solvothermal synthesis and structurally enhanced photocatalytic performance," Advanced Functional Materials, vol. 18, no. 7, pp. 1047–1056, 2008.

5. J. Qu, C. Luo, and Q. Cong, "Synthesis of multi-walled carbon nanotubes/ZnO nanocomposites using absorbent cotton," Nano-Micro Letters, vol. 3, pp. 115–120, 2011.

6.   S. Chu, G. Wang, W. Zhou et al., "Electrically pumped waveguide lasing from ZnO nanowires," Nature Nanotechnology, vol. 6, no. 8, pp. 506–510, 2011.

7.   J. H. Na, M. Kitamura, M. Arita, and Y. Arakawa, "Hybrid p-n junction light-emitting diodes based on sputtered ZnO and organic semiconductors," Applied Physics Letters, vol. 95, no. 25, Article ID 253303, 2009.

8.   P. Sudhagar, R. S. Kumar, J. H. Jung et al., "Facile synthesis of highly branched jacks-like ZnO nanorods and their applications in dye-sensitized solar cells," Materials Research Bulletin, vol. 46, no. 9, pp. 1473–1479, 2011.

9.   Z. L. Wang, R. Yang, J. Zhou et al., "Lateral nanowire/nanobelt based nanogenerators, piezotronics and piezo-phototronics," Materials Science and Engineering R, vol. 70, no. 3–6, pp. 320–329, 2010.

10. J. Xu, J. Han, Y. Zhang, Y. Sun, and B. Xie, "Studies on alcohol sensing mechanism of ZnO based gas sensors," Sensors and Actuators B, vol. 132, no. 1, pp. 334–339, 2008.

11. C. Y. Lu, S. J. Chang, S. P. Chang et al., "Ultraviolet photodetectors with ZnO nanowires prepared on ZnO:Ga/glass templates," Applied Physics Letters, vol. 89, no. 15, Article ID 153101, 2006.

12. S. Cho, S. Kim, J. W. Jang, et al., "Large-scale fabrication of sub-20-nm-diameter ZnO nanorod arrays at room temperature and their photocatalytic activity," Journal of Physical Chemistry C, vol. 113, no. 24, pp. 10452–10458, 2009.

13. S. S. Srinivasan, J. Wade, E. K. Stefanakos, and Y. Goswami, "Synergistic effects of sulfation and co-doping on the visible light photocatalysis of TiO_2," Journal of Alloys and Compounds, vol. 424, no. 1-2, pp. 322–326, 2006.

14. H. Zhang, X. Lv, Y. Li, Y. Wang, and J. Li, "P25-graphene composite as a high performance photocatalyst," ACS Nano, vol. 4, no. 1, pp. 380–386, 2010.

15. D. Y. Goswami, "Decontamination of ventilation systems using photocatalytic air cleaning technology,"Journal of Solar Energy Engineering, Transactions of the ASME, vol. 125, no. 3, pp. 359–365, 2003.

16. A. Sapkota, A. J. Anceno, S. Baruah, O. V. Shipin, and J. Dutta, "Zinc oxide nanorod mediated visible light photoinactivation of model microbes in water," Nanotechnology, vol. 22, no. 21, Article ID 215703, 2011.

17. M. Ladanov, M. K. Ram, G. Matthews, and A. Kumar, "Structure and opto-electrochemical properties of ZnO nanowires grown on n-Si substrate," Langmuir, vol. 27, no. 14, pp. 9012–9017, 2011.

18. D. M. Fouad and M. B. Mohamed, "Comparative study of the photocatalytic activity of semiconductor nanostructures and their hybrid metal nanocomposites on the photodegradation of malathion," Journal of Nanomaterials, vol. 2012, Article ID 524123, 8 pages, 2012.

19. A. A. Khodja, T. Sehili, J. F. Pilichowski, and P. Boule, "Photocatalytic degradation of 2-phenylphenol on TiO_2 and ZnO in aqueous suspensions," Journal of Photochemistry and Photobiology A, vol. 141, no. 2-3, pp. 231–239, 2001.

20. G. Marcì, V. Augugliaro, M. J. López-Muñoz et al., "Preparation characterization and photocatalytic activity of polycrystalline ZnO/TiO_2 systems. 2. Surface, bulk characterization, and 4-nitrophenol photodegradation in liquid-solid regime," Journal of Physical Chemistry B, vol. 105, no. 5, pp. 1033–1040, 2001

21. N. Sobana and M. Swaminathan, "The effect of operational parameters on the photocatalytic degradation of acid red 18 by ZnO," Separation and Purification Technology, vol. 56, no. 1, pp. 101–107, 2007.

22. Q. Wan, T. H. Wang, and J. C. Zhao, "Enhanced photocatalytic activity of ZnO nanotetrapods," Applied Physics Letters, vol. 87, no. 8, pp. 1–3, 2005.

23. N. V. Kaneva, D. T. Dimitrov, and C. D. Dushkin, "Effect of nickel doping on the photocatalytic activity of ZnO thin films under UV and visible light," Applied Surface Science, vol. 257, no. 18, pp. 8113–8120, 2011.

24. S. Sakthivel, B. Neppolian, M. V. Shankar, B. Arabindoo, M. Palanichamy, and V. Murugesan, "Solar photocatalytic degradation of azo dye: comparison of photocatalytic efficiency of ZnO and TiO_2," Solar Energy Materials and Solar Cells, vol. 77, no. 1, pp. 65–82, 2003.

25. S. Ahmed, M. G. Rasul, W. N. Martens, R. Brown, and M. A. Hashib, "Heterogeneous photocatalytic degradation of phenols in wastewater: a review on current status and developments," Desalination, vol. 261, no. 1-2, pp. 3–18, 2010.

26. H. Ghayour, H. R. Rezaie, S. Mirdamadi, and A. A. Nourbakhsh, "The effect of seed layer thickness on alignment and morphology of ZnO nanorods," Vacuum, vol. 86, no. 1, pp. 101–105, 2011.

27. W. Y. Wu, C. C. Yeh, and J. M. Ting, "Effects of seed layer characteristics on the synthesis of ZnO nanowires," Journal of the American Ceramic Society, vol. 92, no. 11, pp. 2718–2723, 2009.

28. N. V. Suramwar, S. R. Thakare, N. N. Karade, and N. T. Khati, "Green synthesis of predominant (111) facet CuO nanoparticles: heterogeneousand recyclable catalyst for N-arylation of indoles," Journal of Molecular Catalysis A, vol. 359, pp. 28–34, 2012.

29. G. Kenanakis, D. Vernardou, E. Koudoumas, and N. Katsarakis, "Growth of c-axis oriented ZnO nanowires from aqueous solution: the decisive role of a seed layer for controlling the wires' diameter,"Journal of Crystal Growth, vol. 311, no. 23-24, pp. 4799–4804, 2009.

30. A. Sugunan, V. K. Guduru, A. Uheida, M. S. Toprak, and M. Muhammed, "Radially oriented ZnO nanowires on flexible poly-L-lactide nanofibers for continuous-flow photocatalytic water purification,"Journal of the American Ceramic Society, vol. 93, no. 11, pp. 3740–3744, 2010.

31. G. Kenanakis and N. Katsarakis, "Light-induced photocatalytic degradation of stearic acid by c-axis oriented ZnO nanowires," Applied Catalysis A, vol. 378, no. 2, pp. 227–233, 2010.

32. Y. Zhang, J. Xu, P. Xu, Y. Zhu, X. Chen, and W. Yu, "Decoration of ZnO nanowires with Pt nanoparticles and their improved gas sensing and photocatalytic performance," Nanotechnology, vol. 21, no. 28, Article ID 285501, 2010.

33. C. Ma, Z. Zhou, H. Wei, Z. Yang, Z. Wang, and Y. Zhang, "Rapid large-scale preparation of ZnO nanowires for photocatalytic application," Nanoscale Research Letters, vol. 6, no. 1, article 536, 2011.

34. S. Baruah, M. Abbas, M. Myint, T. Bora, and J. Dutta, "Enhanced visible light photocatalysis through fast crystallization of zinc oxide nanorods," Beilstein Journal of Nanotechnology, vol. 1, pp. 14–20, 2010.

Chapter 7

Influence of Crystal Structure of Nanosized ZrO$_2$ on Photocatalytic Degradation of Methyl Orange

Sulaiman N Basahel[1], Tarek T Ali1[2], Mohamed Mokhtar[1, 3], and Katabathini Narasimharao[1]

[1]Department of Chemistry, Faculty of Science, King Abdulaziz University, Jeddah 21589, Kingdom of Saudi Arabia
[2]Chemistry Department, Faculty of Science, Sohag University, Sohag 82524, Egypt
[3]Physical Chemistry Department, National Research Centre, El Buhouth St., Dokki, Cairo 12622, Egypt

ABSTRACT

Nanosized ZrO$_2$ powders with near pure monoclinic, tetragonal, and cubic structures synthesized by various methods were used as catalysts for photocatalytic degradation of methyl orange. The structural and textural properties of the samples were analyzed by X-ray diffraction, Raman spectroscopy, TEM, UV-vis, X-ray photoelectron spectroscopy (XPS), and N_2 adsorption measurements. The performance of synthesized ZrO_2 nanoparticles in the photocatalytic degradation of methyl orange under UV light irradiation was evaluated. The photocatalytic activity of the pure monoclinic ZrO_2 sample is higher than that of the tetragonal and cubic ZrO_2 samples under optimum identical conditions. The characterization results revealed that monoclinic ZrO_2 nanoparticles possessed high crystallinity and mesopores with diameter of 100 Å. The higher activity of the monoclinic ZrO_2 sample for the photocatalytic degradation of methyl orange can be attributed to the combining effects of factors including the presence of small amount of oxygen-deficient zirconium oxide phase, high crystallinity, large pores, and high density of surface hydroxyl groups.

BACKGROUND

The rapid growth of the textile industry has led to the accumulation of various organic pollutants, with dyes accumulating in bodies of water as a particularly severe example. This type of aquatic pollution has indirect or direct adverse effects on the biosphere [1]. Photocatalysis is one promising approach to protect the aquatic environment based on its ability to oxidize low concentrations of organic pollutants in water [2],[3]. In the past two decades, many oxide and sulfide semiconductors such as TiO_2, ZnO, WO_3, $SrTiO_3$, ZnS, and CdS were applied as photocatalysts for environmental control technology and also a wide range of chemical reactions [4]. Recently, Kuriakose et al. [5],[6], Cheng et al. [7], and Ren et al. [8] successfully employed ZnO- and TiO_2-based nanomaterials for photocatalytic degradation of organic dyes. ZrO_2 has been considered as a photocatalyst in different chemical reactions due to its relatively wide band gap value E_g and the high negative value of the conduction band potential [9]. The reported band gap energy of ZrO_2 range was between 3.25 and 5.1 eV, depending on the preparation technique of the sample [10].

It is reported that a good manipulation of ZrO_2 morphological tuning, porous structure control, and crystallinity development is required in order to enhance the light harvesting capability, prolong the lifetime of photoinduced electron-hole pairs, and facilitate the reactant accessibility to surface active sites [11]. As ZrO_2 is used in a wide variety of applications in addition to photocatalysis, the fabrication of identical ZrO_2 nanoscale structures has been recently attracted a great deal of interest. Nanocrystalline ZrO_2 with various attractive morphologies has been effectively prepared by different synthesis methods like hydrothermal synthesis, sol-gel synthesis, precipitation, and thermal decomposition [12].

It is well known that ZrO_2 has three polymorphs [13]: monoclinic, tetragonal, and cubic. Preparation methods play an important role in determining the final crystal structure of ZrO_2. Although the different surface properties on different ZrO_2 polymorphs have been extensively studied [14],[15], the effect of crystal structures on photocatalysis has rarely been investigated.

Nawale et al. [16] synthesized ZrO_2 samples using thermal plasma reactor at different operating pressures. The sample which contained both tetragonal and monoclinic phases synthesized at 1.33 bar of operating pressure showed the highest photocatalytic activity. The presence of tetragonal phase along with monoclinic phase indicates the crystallographic rearrangement in ZrO_2 due to the oxygen vacancies. The authors related the photocatalytic properties of ZrO_2 with the trap levels present in it due to oxygen vacancies. It was observed that the photocatalytic response tracks the energy gap of the monoclinic phase which varies with the varying synthesis parameters.

Zhao et al. [17] used anodization method to synthesize ZrO_2 nanotubes with a length of 25 μm, inner diameter of 80 nm, and wall thickness of 35 nm. The authors observed 97.6 decolorization percentage of methyl orange in 8 h at optimal pH value 2. Ismail et al. [18] synthesized 6-μm-thick anodic oxide film with nanotubular ZrO_2 structure, and the authors tested the photocatalytic ability of the ZrO_2 nanotubes. The authors reported 30% of methyl orange degradation under UV light in the presence of the cubic/tetragonal ZrO_2 nanotubes after 120 min of reaction.

Jiang et al. [19] used zirconium foil to anodize in electrolyte containing 1 M $(NH_4)_2SO_4$ and 0.25 wt.% NH_4F to in situ construct the ZrO_2 nanotubes on the surface. The authors reported that ZrO_2 nanotubes showed excellent photocatalytic performance with methyl orange photodegradation rate of 94.4% after 240 min. They also claimed that photocatalysis performance was due to the hydroxyl group absorbing on the surface.

Shu et al. [20] synthesized tetragonal star-like ZrO_2 nanostructures using hydrothermal synthesis method. The authors used ZrO_2 nanostructures for the photodegradation of anionic dyes including methyl orange, in acidic, neutral, and weak basic aqueous solutions. They observed that the ZrO_2 sample offered complete degradation of methyl orange within 60 min; however, authors have not studied the stability and reusability of the synthesized ZrO_2 nanomaterial.

The objective of the present study is to synthesize nanocrystalline mesoporous monoclinic, tetragonal, and cubic ZrO_2 samples with high surface area using fairly simple experimental procedures. In this work, nanosized pure monoclinic, tetragonal, and cubic ZrO_2 samples were prepared and the physico-chemical properties of the samples were performed by different characterization techniques. The photocatalytic degradation of methyl orange over the three ZrO_2 samples were studied and correlated to the phase structure, specific surface area, and electronic properties of the catalysts.

METHODS

Materials

Zirconyl chloride, zirconium isopropoxide, sodium hydroxide solution, methyl orange, and hydrochloric acid were purchased from Aldrich, Dorset, England, UK. All chemicals used in this study were analytical grade and used directly without further purification. Deionzied water was used for the preparation of the methyl orange standard solution as well as the respective dilutions.

Synthesis of Pure Monoclinic, Tetragonal, and Cubic ZrO$_2$ Samples

Monoclinic ZrO$_2$

A near pure monoclinic nanocrystalline ZrO$_2$ was synthesized by following the method reported by Guo et al. [21]. The zirconyl chloride was dissolved in deionized water so that the final concentration of zirconium was 38.7 g per liter (0.42 M). Of the zirconyl chloride solution, 15 ml was added to 300 mL deionized water in a glass beaker, and then concentrated aqueous ammonia was added rapidly to the solution with constant stirring until pH 4.5. The resultant precipitate was aged in the mother liquor for 24 h. After filtration, it was washed several times with dilute ammonia and hot deionized water (80°C) until chloride ions were no longer detectable in the washing water (AgNO$_3$ test) and then dried at 100°C for 12 h. The synthesized sample was calcined at 500°C for 3 h in air with a ramp rate of 1°C min^{-1} and kept isothermally for 3 h and was annotated as m-ZrO$_2$.

Tetragonal ZrO$_2$

Pure tetragonal ZrO$_2$ was synthesized by the following reported procedure in the literature [22]. Zirconium oxychloride and ammonia solution (25% w/w) solutions were prepared using deionized water. First, 50 mL of 2.5 M ammonia solution was added to 50 mL of 0.1 M zirconium oxychloride solution drop by drop in a beaker and the mixture was stirred vigorously at room temperature for 4 h. The white zirconium hydroxide precipitates in time of addition of ammonia solution. The obtained precipitate was separated by centrifugation at 4,000 rpm, washed with water and ethanol for several times. Then, the precipitate was transferred into Teflon-lined autoclave, and the autoclave was kept at 100°C for 12 h. Finally, the white powder was calcined in furnace at 500°C for 3 h with a ramp rate of 1°C min^{-1} and kept at this temperature for 3 h and was annotated as t-ZrO$_2$.

Cubic ZrO₂

Cubic ZrO_2 was synthesized by hydrothermal method reported my Tahir et al. [23]. In a Teflon vessel, 1 g of zirconium isopropoxide was dissolved in 6 mL of ethanol (99.8%) and then the Teflon vessel was kept in a desiccator containing a Petri dish filled with water at the bottom. The diffusion experiment was stopped after 12 h, followed by the addition of 25 mL of 10 M NaOH aqueous solution. Then, the reaction vessel was sealed into a stainless steel hydrothermal bomb, which was heated to 180°C for 18 h. After the autoclave was cooled down to room temperature, the products were filtered and repeatedly washed with 0.1 M HNO_3, 1 N HCl, and deionized water. After drying under vacuum for 3 h, a white soft and fibrous powder was obtained. The obtained powder was calcined at 500°C for 3 h in air with a ramp rate of 1°C min^{-1} and kept isothermally for 3 h and was annotated as c-ZrO_2.

Material Characterization

X-ray powder diffraction (XRD) studies were performed for all of the prepared solid samples using a Bruker diffractometer (Bruker D8 advance target; Bruker AXS, GmbH, Karlsruhe, Germany). The patterns were run with copper $K\alpha_1$ and a monochromator ($\lambda = 1.5405$ Å) at 40 kV and 40 mA. The crystallite size of the ZrO_2 was calculated using Scherrer's equation;

$$D = B\lambda / \beta_{1/2}\cos\theta$$

(1)

where D is the average crystallite size of the phase under investigation, B is the Scherer constant (0.89), λ is the wavelength of the X-ray beam used (1.54056 A°), $\beta_{1/2}$ is the full width at half maximum (FWHM) of the diffraction peak, and θ is the diffraction angle. The identification of different crystalline phases in the samples was performed by comparing the data with the Joint Committee for Powder Diffraction Standards (JCPDS) files.

The Raman spectra of the samples were measured with a Bruker Equinox 55 FT-IR spectrometer equipped with an FRA106/S FT Raman module and a liquid N_2-cooled Ge detector using the 1,064-nm line of a Nd:YAG laser with an output laser power of 200 mW.

A Philips CM200FEG microscope (Philips, Amsterdam, The Netherlands), 200 kV, equipped with a field emission gun was used for HRTEM analysis. The coefficient of spherical aberration was Cs = 1.35 mm. The information limit was better than 0.18 nm. High-resolution images with a pixel size of 0.044 nm were taken with a CCD camera.

The textural properties of the prepared samples were determined from nitrogen adsorption/desorption isotherm measurements at −196°C using Autosorb automated gas sorption system (Quantachrome, Boynton Beach, FL, USA). Prior to measurement, each sample was degassed for 6 h at 150°C. The specific surface area, S_{BET}, was calculated by applying the Brunauer-Emmett-Teller (BET) equation. The average pore radius was estimated from the relation $2V_p/S_{BET}$, where V_p is the total pore volume (at $P/P^0 = 0.975$). Pore size distribution over the mesopore range was generated by the Barrett-Joyner-Halenda (BJH) analysis of the desorption branches, and the values for the average pore size were calculated.

The X-ray photoelectron spectroscopy (XPS) measurements were carried out by using a SPECS GmbH X-ray photoelectron spectrometer (SPECS, Berlin, Germany). Prior to analysis, the samples were degassed under vacuum inside the load lock for 16 h. The binding energy of the adventitious carbon (C 1 s) line at 284.6 eV was used for calibration, and the positions of other peaks were corrected according to the position of the C 1 s signal. For the measurements of high-resolution spectra, the analyzer was set to the large area lens mode with energy steps of 25 m eV and in fixed analyzer transmission (FAT) mode with pass energies of 34 eV and dwell times of 100 ms. The photoelectron spectra of the four samples were recorded with the acceptance area and angle of 5 mm in diameter and up to ±5°, respectively. The base pressure during all measurements was 5×10^{-9} mbar. A standard dual anode excitation source with Mg K_α (1,253.6 eV) radiation was used at 13 kV and 100 W.

The UV-vis absorption spectra in transmittance mode were recorded on a Thermo Scientific (Evolution 600 UV-vis; Thermo Fisher Scientific, Waltham, MA, USA) instrument. The optical bandgap of the samples is measured by plotting a $(\alpha h\upsilon)^2$ versus $h\upsilon$. The extrapolation of the straight line in the graph to $(\alpha h\upsilon)^2 = 0$ gives the value of the energy band gap.

Photocatalytic Degradation of Methyl Orange

Photocatalytic activity measurements were carried out in a homebuilt reactor. The reactor is a wooden box with dimensions of 100 cm height, 100 cm width, and 60 cm thickness, equipped with a 12-V transformer for an electric exhaust fan. Six 18 W UV lamps (60 cm × 2.5 cm) of approximately 350 to 400 nm (F20 T8 BLB) were used; the total power of the UV light at the surface of the test suspension measured with a Newport 918DUVOD3 detector (Newport Corporation, Irvine, CA, USA) and power meter was 13 W/m². In a typical experiment, 100 mL of aqueous methyl orange solution (10 mg/L) was stirred (300 rpm) with 100 mg of the different photocatalysts. The resulting suspension was equilibrated by stirring for 1 h to stabilize the absorption of methyl orange dye over the surface of catalyst before exposing to the UV light. Samples were withdrawn at 10 min intervals, filtered through a 0.2-mm PTFE Millipore membrane filter (Millipore, Billerica, MA, USA) to remove suspended catalyst agglomerates, and finally analyzed using the UV-vis spectrometer (Thermo Fisher Scientific Evolution 160) in the range between 250 and 600 nm. The decolorization rate percentages of methyl orange were calculated by the following equation:

$$\text{Decolorization\%} = \left(1 - \frac{C}{C_o}\right) \times 100$$

(2)

where C_o is the concentration of methyl orange before illumination and C is the concentration after a certain irradiation time.

RESULTS AND DISCUSSION

X-Ray Powder Diffraction

The XRD patterns of synthesized m-ZrO_2, t-ZrO_2, and c-ZrO_2 samples and corresponding JCPDS reference patterns are shown in Figure 1. XRD pattern of the m-ZrO_2 sample showed intensive diffraction patterns at $2\theta = 24.2°$, $28.2°$, $31.4°$, and $34.3°$ which are corresponding to monoclinic ZrO_2 crystal phase [JCPDS 37-1484]. It is observed that there is one major peak at $2\theta = 25.4°$ and another small peak at $22°$, which are not indexed for monoclinic ZrO_2 phase. These peaks can be indexed to the oxygen-deficient zirconium oxide, $ZrO_{0.35}$ phase [JCPDS; 17-0385, hexagonal, space group P6322]. To determine the purity of monoclinic phase of the m-ZrO_2 sample, volume percent of monoclinic and oxygen-deficient zirconium oxide phase present in the m-ZrO_2 sample was determined from the integrated intensities of the diffraction peaks (−111) (111) of m-ZrO_2 at $2\theta = 28.5°$ and $31.5°$, respectively, and the diffraction line (101) of oxygen deficient ZrO_2 at $2\theta = 25.4°$. We used the expressions (3) and (4) reported in the literature [24].

$$\%M_{monoclinic} = \varSigma\, M_{monoclinic} \times 100 \,/\, OD_{oxygen\ deficient} + \varSigma\, M_{monoclinic} \tag{3}$$

$$\%OD_{oxygen\ deficient} = 100 - \%M_{monoclinic} \tag{4}$$

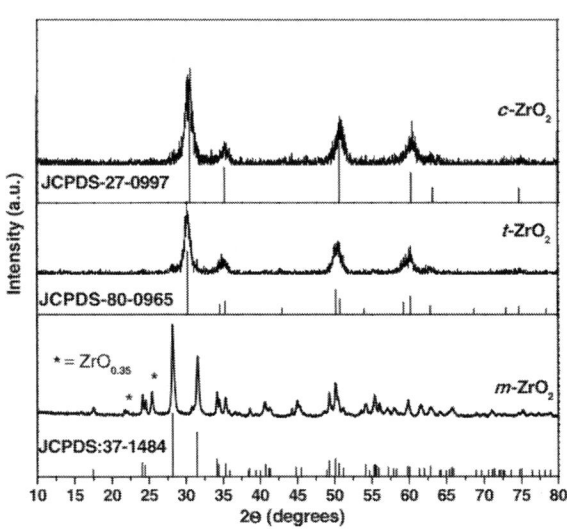

Figure 1: XRD patterns of the different zirconia samples.

The percentages of monoclinic and oxygen deficient phases were found to be 97% and 3%, respectively, for the m-ZrO$_2$ sample. There are no additional unindexed peaks in the m-ZrO$_2$ sample.

XRD pattern of the t-ZrO$_2$ sample showed peaks for pure tetragonal phase of ZrO$_2$ [JCPDS 80-0965] at $2\theta = 30.2°$, 35.2°, 50.6°, and 60.2°. No additional peaks corresponding to any other phase was observed in XRD pattern of this sample. All of the diffraction peaks of the XRD pattern of the c-ZrO$_2$ sample can be indexed to the standard pattern of the pure cubic phase of ZrO$_2$. Peaks at $2\theta = 30.3°$, 35.14°, 50.48°, and 60.2° reveal the presence of (111), (200), (220), and (311) planes, respectively, of cubic ZrO$_2$ according to JCPDS CAS number 27-0997. These observation indicates that the m-ZrO$_2$ sample is near pure monoclinic; however, the t-ZrO$_2$ and c-ZrO$_2$ samples did not show presence of any additional phases or impurities indicating that these two phases are pure in composition. In addition, the intensities of diffraction peaks of the m-ZrO$_2$ sample were much higher than those of the t-ZrO$_2$ and c-ZrO$_2$ samples indicating that the m-ZrO$_2$ sample is highly crystalline than the t-ZrO$_2$ and c-ZrO$_2$ samples.

However, it is known that assignment of cubic and tetragonal structures, based solely on the X-ray diffraction analysis, can be misleading because the cubic and tetragonal structures ($a_0 = 0.5124$ nm for cubic and $a_0 = 0.5094$ nm and $c_0 = 0.5177$ nm for tetragonal structures) are very similar [25]. The authors also reported that the tetragonal structure can be distinguished from the cubic structure by the presence of the characteristic splitting of the diffraction peaks, whereas the cubic phase exhibits only single peaks. A significant line broadening obscured any clear distinction between the tetragonal and cubic polymorphs of ZrO$_2$ (Figure 1). However, this detection is possible by measuring with high step counting times the (112) Bragg reflection of the tetragonal structure, which is forbidden in the cubic symmetry [26]. As it can be observed in Figure 1, a shift of the peak positions to higher 2θ values occurred. This shift may indicate a decrease in the lattice parameters.

The crystallite size was calculated using the Scherrer›s equation (1). The average crystallite sizes of the monoclinic phase, calculated from the (111) diffraction peak was found to be 34 nm. Similarly, the average crystallite sizes, calculated from the (111) diffraction peak of the tetragonal and cubic phases, were found to be 17 and 20 nm for the t-ZrO$_2$ and c-ZrO$_2$ samples, respectively.

Raman Spectroscopy

In order to confirm the crystalline structure of the samples, the Raman spectra of the samples were obtained and shown in Figure 2. From this figure, we can see that the m-ZrO$_2$ sample showed several peaks centering at 183, 301, 335, 381, 476, 536, 559, 613, and 636 cm^{-1}. The strong peaks are at 183, 335, and 476 cm^{-1}. The exhibited bands are clearly indicating that the m-ZrO$_2$ sample possessed dominant monoclinic phase of ZrO$_2$[27]. The t-ZrO$_2$ sample showed peaks at 149, 224, 292, 324, 407, 456, and 636 cm^{-1}, and the peak positions are in quite accordance with the reported values for tetragonal phase of ZrO$_2$[28].

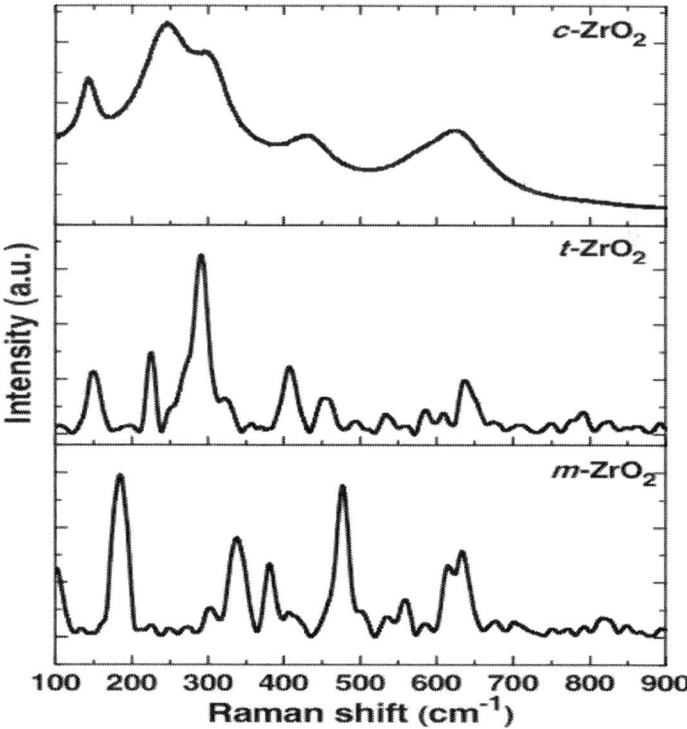

Figure 2: Raman spectra of different zirconia samples.

The Raman spectrum for c-ZrO$_2$ is characterized by a narrow band at 145 cm^{-1} and broad bands centered at 246, 301, 436, and 625 cm^{-1}. Gazzoli et al. [29] reported that the Raman peak at 149 cm^{-1} is common for both of tetragonal and cubic phases, and cubic ZrO$_2$ presents the strong band between 607 and 617 cm^{-1}. The c-ZrO$_2$ sample in this study clearly showed the broad peak centered at 625 cm^{-1}, and this sample also shows poorly defined features related to the disordered oxygen sub-lattice whereas tetragonal ZrO$_2$ exhibits several well-defined sharp bands because of the symmetry reduction [30]. In addition, the highly intense peaks at 292 and 636 cm^{-1} which are main characteristic bands of tetragonal ZrO$_2$ cannot be found in the spectrum of the c-ZrO$_2$ sample, which indicates the absence of tetragonal ZrO$_2$ phase in this sample. Kontoyannis et al. [31] also reported that cubic ZrO$_2$ shows amorphous-like Raman spectrum with one broad band at 530 to 670 cm^{-1}. The features of Raman spectrum of c-ZrO$_2$ as shown in Figure 2 is in accordance with the spectral results reported in the literature.

Transmission Electron Microscopy

The TEM images for the m-ZrO$_2$, t-ZrO$_2$, and c-ZrO$_2$ samples are shown in Figure 3A,B,C, respectively. Tightly packed dumbbell-shaped particles can be observed in the low magnification TEM images of three samples. The average particle size for the m-ZrO$_2$, t-ZrO$_2$, and c-ZrO$_2$ samples was found to be 24, 18, and 8 nm, respectively. There are conflicting reports in the literature regarding the phase structure of ZrO$_2$ particles in smaller size (less than 10 nm). Some authors reported that cubic ZrO$_2$ phase exists as fine nanoparticles [32], and few other researchers reported that tetragonal ZrO$_2$ phase exists in smaller size than cubic phase [33]. However, in the present work, it is clear that in the c-ZrO$_2$ sample, pure cubic ZrO$_2$ phase possessed smaller particles size than the t-ZrO$_2$ sample (pure tetragonal).

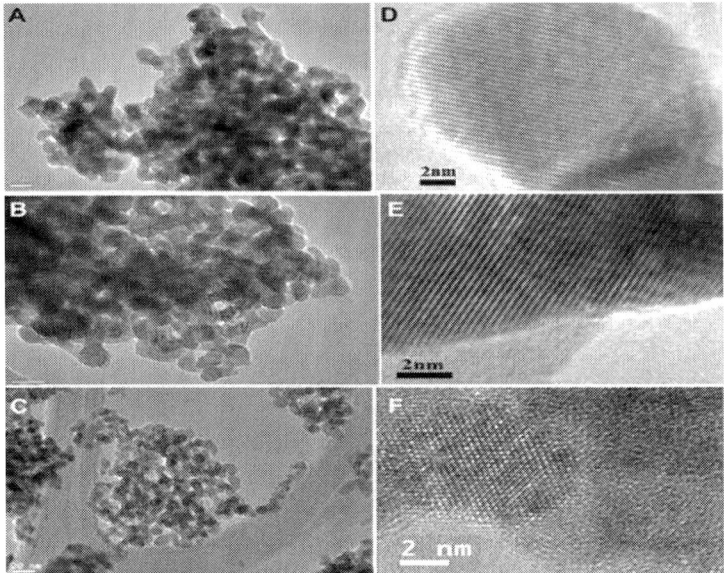

Figure 3: TEM images of (A) m-ZrO$_2$(B)t-ZrO$_2$(C)c-ZrO$_2$and HRTEM images of (D)m-ZrO$_2$(E)t-ZrO$_2$(F)c-ZrO$_2$.

High-resolution Transmission Electron Microscopy

In order to authenticate the ZrO$_2$ phase existed in the samples, high-resolution transmission electron microscopy (HRTEM) was carried out on particles of the three samples. Figure 3D represents the HRTEM image of the m-ZrO$_2$ sample. The image clearly showed well-resolved lattice fringes. The distance between the fringes was calculated to be 0.297 nm which can be attributed to the interplanar spacing corresponding to (111) plane of monoclinic ZrO$_2$[34]. The HRTEM image of t-ZrO$_2$was shown in Figure 3E. This image also showed well-resolved equidistant lattice fringes. The distance between the parallel fringes was calculated to be 0.296 nm which can be attributed to the well-recognized lattice d-spacing of (111) plane of tetragonal ZrO$_2$[35]. A typical HRTEM image of particles of the c-ZrO$_2$ sample is shown in Figure 3F. The image shows equidistant parallel fringes which depict single crystalline nature of the particle. The distance between the parallel fringes was calculated to be 0.291 nm which is the well-recognized lattice d-spacing of (111) plane of cubic ZrO_2 [36].

BET Surface Area

A typical nitrogen adsorption-desorption isotherms of the samples are shown in Figure 4. The adsorption-desorption patterns of the three ZrO_2 samples belong to the typical IUPAC IV-type with the H2-type hysteresis loop, which is a characteristic of particles with uniform size and mesoporous structure [37]. From the figure, it is clear that all the three samples showed type IV isotherms with hysteresis loop at $P/P°$ = 0.45 to 0.95. However, each sample exhibited a different type of hysteresis loop suggesting that pore size and shape were not same in these samples. The H2-type adsorption hysteresis can be explained as a consequence of the interconnectivity of pores. It was reported that in such systems, the distribution of pore sizes and pore shapes are not well defined or irregular. A sharp step on desorption isotherm is usually understood as a sign of interconnection of the pores. The shape of hysteresis loop of the m-ZrO_2 sample suggesting that this sample possessed pores known as 'ink-bottle' type [38].

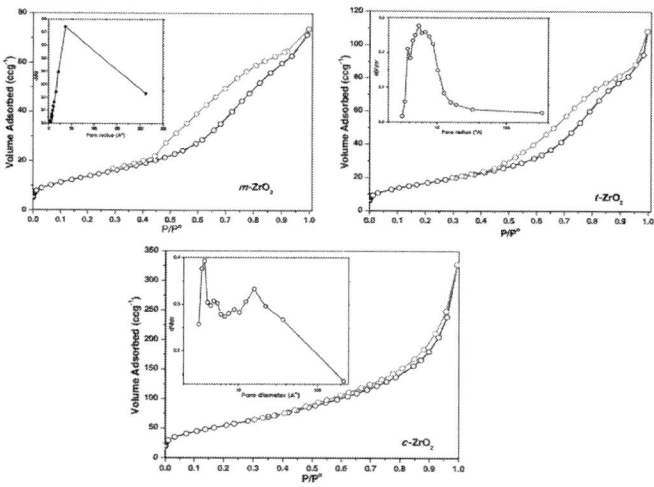

Figure 4: Nitrogen adsorption-desorption isotherms and pore size distribution (inset) of the samples.

The pore size distribution patterns of the synthesized ZrO_2 samples were shown in the inset of Figure 4. Pore size distribution of the m-ZrO_2 sample reveals a broad and monomodal distribution of the pore dimension in the mesoporous region. In addition, the c-ZrO_2 sample showed a narrower pore size distribution than t-ZrO_2. The porosity of the m-ZrO_2 and t-ZrO_2 samples appears to be arising from non-crystalline intra-aggregate voids and spaces formed as a result of inter-particle contact [39]. The narrow and broad bimodal distribution of pores can be observed in the case of thec-ZrO_2 sample. The porosity of this sample appears to be framework porosity which corresponds to the porosity within the uniform channels of ZrO_2 structure.

The textural properties of the synthesized ZrO_2 samples from the adsorption-desorption data was tabulated in Table 1. Specific surface area of the m-ZrO_2 sample is 65 m^2g^{-1}, an average pore radius of 50 Å, and a total pore volume of 0.626 cm^3g^{-1}. The t-ZrO_2 and c-ZrO_2 samples possessed the surface area of 74 and 204 m^2g^{-1}; the drastic increase of the surface area of the c-ZrO_2 sample could be due to very small size of particles (TEM results). However, the c-ZrO_2 sample possessed pores with small radius (19 Å) than the t-ZrO_2 (28.3 Å) and m-ZrO_2 (50 Å) samples.

Table 1: Textural properties of the catalysts from N2adsorption measurements

Catalyst	SBET(m2g−1)	Vp(cm3g−1)	Pore radius (Å)
m-ZrO2	65	0.626	50.0
t-ZrO2	74	0.521	28.3
c-ZrO2	204	0.508	19.0

Basahel et al.

Basahel et al. Nanoscale Research Letters 2015 10:73   doi:10.1186/s11671-015-0780-z

X-ray Photoelectron Spectroscopy

It is known that XPS is a very sensitive tool in analyzing the chemical state of Zr cations in ZrO_2 and its composites [40]. Figure 5A,B displays the XPS spectra of the Zr 3d and O 1 s core levels of the three samples, respectively. The peaks located at 181.3 and 183.8 eV are attributed to the spin-orbit splitting of the Zr 3d components, Zr 3d5/2 and Zr 3d3/2. The binding energy of O 1 s in ZrO_2 is located at 530.1 eV.

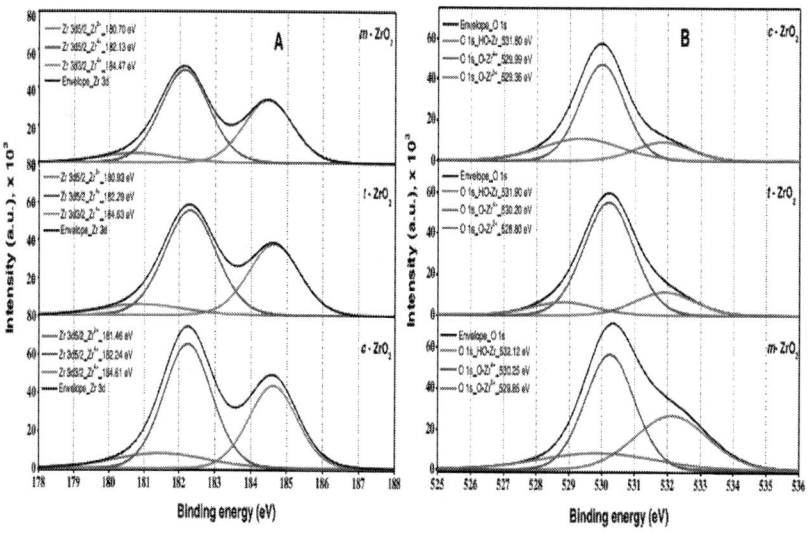

Figure 5: Deconvoluted XPS spectra of samples (A) Zr (3dand5d) (B) O (1 s).

Deconvolution of the spectra produces peaks attributed to the existence of two kinds of zirconium species, referred as Zr^{2+} species with low binding energy in the range 180.7 to 181.4 eV and Zr^{4+}species with higher binding energy in the range of 182.1 to 182.3 eV. It should be noted that the fraction of Zr^{4+} species for all samples is larger compared to that of species Zr^{2+}. It is reported that the binding energy of Zr^{4+} species in pure ZrO_2 is around 182.6 eV [41]; however, slightly lower values compared to that of stoichiometric ZrO_2 were observed especially for the sample m-ZrO_2(182.1 eV), probably, due to some oxygen deficiency. The position shift toward the lower binding energy might be associated with the holes created by oxygen vacancies in the ZrO_2 lattice [42].

Kawasaki [43] reported that Zr 3d components, Zr 3d5/2 and Zr 3d3/2 for cubic ZrO_2 can be observed at 182.0 and 184.4 eV, respectively. The same components for tetragonal ZrO_2 and monoclinic ZrO_2samples appeared at 182.7 and 184.7 eV [44] and 182.2 eV and 184.6 eV [45], respectively. The binding energy values of Zr 3d components observed in this study are in accordance with values reported in the literature.

The O 1 s broad peaks can be deconvulated into three peaks at the corresponding position using XPS Casa Software, whose relative contents are shown in Table 2. Navio et al. [9] observed two types of oxygen species in the ZrO_2 sample, oxygen species of ZrO_2 and oxygen species of Zr-OH, whose binding energy is in the range of 529.8 to 530.3 and 530.9 to 532.2 eV, respectively. It was also reported that the oxygen species with binding energy 531.0 eV are attributed to Zr-OH groups [46]. All the three samples showed XPS peaks corresponding to Zr-OH, Zr^{4+}-O, and Zr^{2+}-O species in different proportions [43]. The m-ZrO_2 sample showed highest 12.3 mass percentage Zr-OH groups bounded to Zr atom, while t-ZrO_2 and c-ZrO_2 have 8% and 6.1%, respectively. These features of the XPS spectra indicate that the c-ZrO_2 and t-ZrO_2 samples were regular surfaces and with no apparent defect relative to the Zr^{4+} species; this is rather significant. In fact, the m-ZrO_2 sample showed surface defects with more surface hydroxyl groups.

Table 2: Surface composition of catalysts from XPS measurements

Catalyst	Zr (mass%)		O (mass%)		
	Zr4+	Zr2+	O-Zr4+	O-Zr2+	Zr-OH
m-ZrO2	52.7	13.0	16.3	5.7	12.3
t-ZrO2	53.9	14.3	17.3	6.5	8.0
c-ZrO2	54.1	15.1	17.6	7.1	6.1

Basahel et al.

Basahel et al. Nanoscale Research Letters 2015 10:73   doi:10.1186/s11671-015-0780-z

Diffuse-Reflectance UV-vis

Figure 6 represents the UV-vis absorption spectra of the three ZrO_2 samples. It is known that all ZrO_2polymorphs are very similar in vibrational structure, and a minor variation in their band frequencies or intensities infers small differences in the Zr^{4+} distribution in Zr-O sites and the oxygen vacancies and other structural defects [47]. Herrera et al. [48] reported that UV-vis spectra of monoclinic Fe-doped ZrO_2

display two bands at around 245 and 320 nm, which are associated with charge transfer transitions.

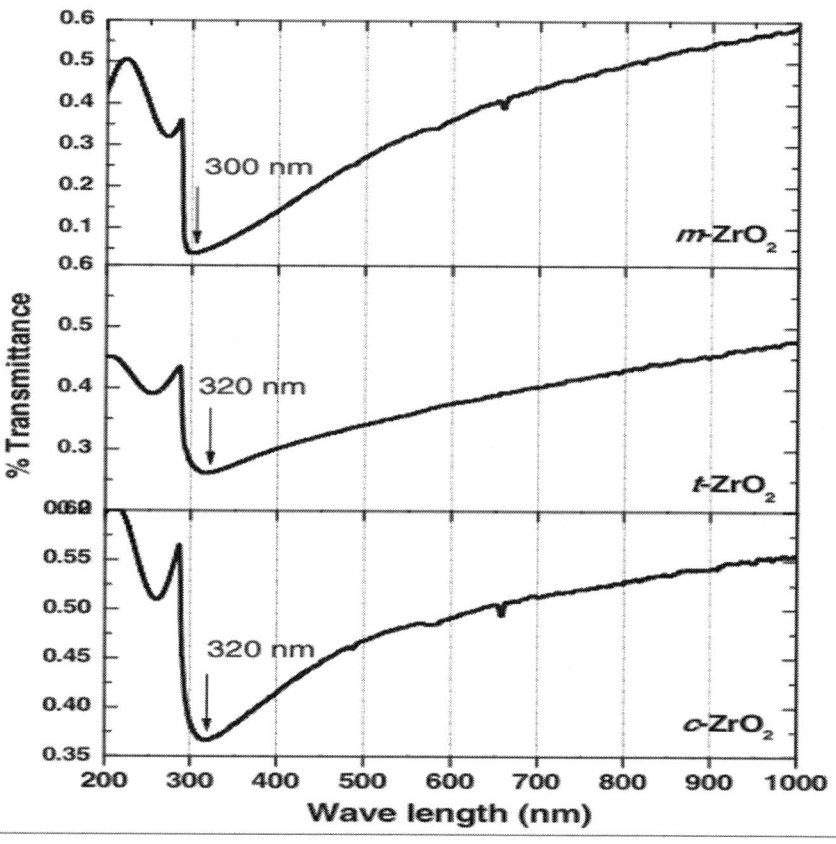

Figure 6: UV-vis absorption spectra of the samples.

Li et al. [49] indicated that the ZrO_2 sample with pure monoclinic ZrO_2 nanoparticles showed a pronounced absorption peak at 270 nm, and tetragonal and cubic ZrO_2 nanoparticles show an absorption peak at 314 nm. The absorptions in the range 250 to 350 nm were assigned to $O_2 \rightarrow Zr^{4+}$ charge transfer transitions with Zr in low coordination states (possibly six) either isolated or present in small Zr_xO_y clusters [10].

Band gap of all the ZrO_2 samples was determined by establishing the relation between $h\upsilon$ and $(\alpha h\upsilon)^2$. The obtained data indicated that the band gap energy for m-ZrO$_2$ (3.25 eV) is lower compared with t-ZrO$_2$ (3.58 eV) and c-ZrO$_2$ (4.33 eV). Crystal structure plays an important role in the electronic structure of ZrO$_2$. This effect is most significant in the d-electron-derived conduction bands (CBs). It is reported that the reduction of the CB gap between the Zr 4d (x2−y2, z2) and the Zr 4d (xy, yz, zx) CBs is present in cubic ZrO$_2$ and disappears in tetragonal ZrO$_2$, and also, substantial volume expansion was observed in the case of monoclinic ZrO$_2$ due to the hybridization of Zr 4d CBs into a new single Zr 4d CB [50].

It was also reported that the pure tetragonal and monoclinic ZrO$_2$ nanoparticles showed energy band gaps of 4.0 and 3.5 eV, respectively [47]. These values are very similar to the values reported in the literature reports. Emeline et al. [51] determined an energy band gap of 5.0 eV for monoclinic ZrO$_2$ thin films calcined at 550°C, and Chang and Doong [52] determined an energy gap of 5.7 eV for the same sample at the same temperature. However, Navio et al. [9] reported an energy band gap of 3.7 eV for monoclinic ZrO$_2$ powders prepared by sol-gel method. These authors claim that the decrease in the band gap energy could be attributed to a highly disordered structure, as a result of the conditions used in the preparation technique. As a consequence of structural defects, some energy levels are introduced into the semiconductor band gap that allow transitions of lower energy and therefore lead to a decrease of the band gap energy.

Photocatalytic Degradation of Methyl Orange

The photocatalytic activity of the three ZrO$_2$ samples was determined by monitoring the degradation of the methyl orange dye. A blank experiment was carried out to confirm that the photo-degradation reaction did not proceed without the presence of either catalyst or the UV radiation. Figure 7 shows the change in the UV-vis absorbance spectra of methyl orange solution (10 ppm) with different irradiation intervals over the ZrO$_2$ samples.

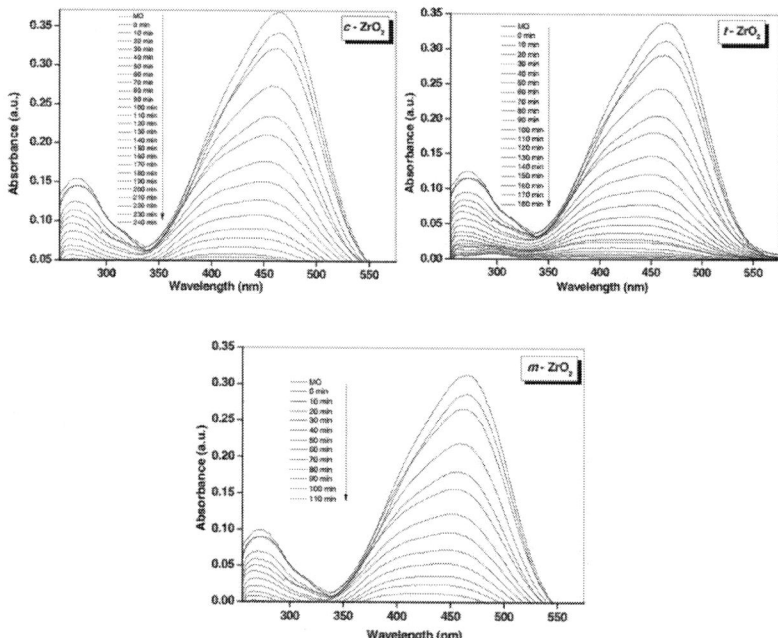

Figure 7: UV-visible absorption changes of methyl orange aqueous solution at 25°C in the presence of the ZrO2 samples.

As mentioned in the experimental section, the catalyst was equilibrated with the methyl orange solution to check for adsorption of the dye on the solid photocatalyst. The spectra depicted in Figure 7 were recorded after the equilibration of the photocatalyst. The UV-vis absorption spectra of methyl orange have a strong characteristic peak at 465 nm and a weak absorption peak at 274 nm.

These absorption peaks become weak and disappear along with the extension of reaction time. The UV-vis results indicate that methyl orange was degraded during the reaction. The decrease in the absorbance of the solution was due to the destruction of the homo- and hetero-polyaromatic rings present in the dye molecules. The m-ZrO$_2$ sample was found to be the most effective catalyst in comparison with the t-ZrO$_2$ and c-ZrO$_2$ samples under identical optimized conditions (Figure 7). Complete degradation of adsorbed dye molecules was observed within 110 min for the m-ZrO$_2$ sample, whereas 180 and 240 min were required for the complete the degradation of adsorbed dye molecules for the t-ZrO_2 and c-ZrO_2 samples under the similar conditions, respectively. The decolorization efficiency of the ZrO_2 samples was calculated using Equation 2. Figure 8 showed degradation efficiency of methyl orange aqueous solution at 25°C in the presence of the three ZrO_2 samples. The m-ZrO_2 photocatalyst showed 99% degradation of methyl orange in 110 min of reaction; however, the t-ZrO_2 and c-ZrO_2 catalysts showed 90% and 80% degradation in the same reaction time, respectively.

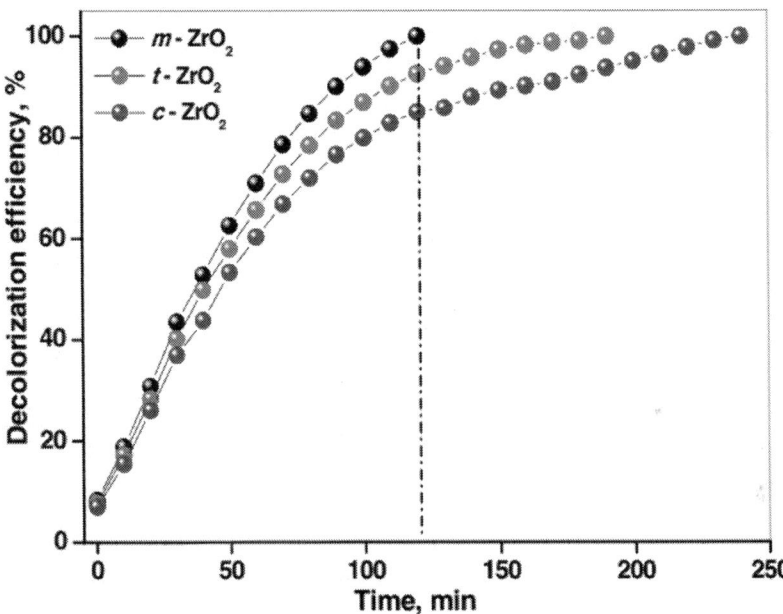

Figure 8: Decolorization efficiency of methyl orange aqueous solution at 25°C in presence of ZrO_2 samples.

It was reported that the structures of nanoscale favor the movement or transfer of electrons and holes generated inside the crystals to the surface [53], which also enhances the photocatalytic activity. The photocatalytic activity of ZrO_2 appears to be strongly dependent on the surface composition. Bachiller-Baeza et al. [54] reported that Lewis acid sites were more abundant on monoclinic ZrO_2 than on tetragonal ZrO_2, and the former brought about stronger surface adsorption sites concerning CO_2 adsorption than the latter. Ma et al. [55] determined and compared the surface properties of ZrO_2 polymorphs. It was found that ZrO_2 polymorphs exhibited different surface hydroxyl and acid-base properties. These differences had great influence on the behavior of CO adsorption and reaction. They also showed that monoclinic ZrO_2 possessed better adsorption properties than the other ZrO_2 structures.

The adsorption capacity of methyl orange per gram of each catalyst was determined under identical conditions. It was observed that m-ZrO_2 (1.95 mg g^{-1}) possessed better adsorption capacity than the t-ZrO_2 (1.05 mg g^{-1}) and c-ZrO_2 (0.65 mg g^{-1}) samples. The photocatalytic activities observed in this study show similar trend that the m-ZrO_2 sample showed high photocatalytic activity even though it possessed less surface area. ZrO_2 nanomaterials have been investigated previously in the photocatalytic degradation of methyl orange; the time period required for the degradation of methyl orange tends to be 120 min or greater. Here, the synthesized pure monoclinic ZrO_2 nanoparticles offered 99% of efficiency in 110 min of reaction time. The observed photocatalytic degradation activity is substantially higher than the activity reported for the ZrO_2 samples in the literature.

It is known that usually high surface area of photocatalyst enhances dye adsorption and subsequent photocatalytic activity. However, it was also reported that the amount of dye adsorption on a catalyst also depends on the adsorption coefficient. Thus, a high adsorption coefficient on a low surface area material could lead to the same amount of adsorbed material per gram of catalyst as low adsorption coefficient on a high surface area material. This could be the reason why the rate (per gram basis) of methyl orange degradation is high for m-ZrO_2 (low surface area) and c-ZrO_2 (high surface area).

The m-ZrO_2 sample possesses the smallest surface area (65 m^{2}g^{-1}) and largest pore size, while thec-ZrO_2 has the largest surface area of 204 m^{2}g^{-1} and smallest pore size. It is interesting that m-ZrO_2 exhibited higher photcatalytic activity than c-ZrO_2. Guo et al. [56] studied the influence of the pore structure of TiO_2 on its photocatalytic performance. They observed that the photocatalytic activity of nanometer TiO_2 is less than that of mesoporous TiO_2.

Shao et al. [57] synthesized ZrO_2-TiO_2 composites which possessed different surface areas and textural properties. Composite samples showed surface area in the range of 270 to 80 m^{2}g^{-1}; however, the sample which has low surface area showed high photocatalytic activity due to it optimum Ti-Zr composition. It is clear from the literature reports that it is possible for catalyst, which possessed low surface area, to offer better photocatalytic activity than the catalyst which possessed high surface area. This is due to the fact that photocatalytic activity can be influenced by several other factors such as crytallinity, composition, particle size distribution, porosity, band gap, and surface hydroxyl density [58].

The process of photocatalytic degradation of methyl orange over ZrO_2 catalysts can be described as follows. The first step involves adsorption of the dye onto the surface of ZrO_2 nanostructure sample. Exposure of dye adsorbed ZrO_2 nanostructures with UV light leads to generation of electron-hole (e^--h^+) pairs in ZrO_2 as indicated in Equation 5. The photogenerated electrons in the conduction band of ZrO_2 interact with the oxygen molecules adsorbed on ZrO_2 to form superoxide anion radicals ($*O_2^-$) (Equation 6). The holes generated in the valence band of ZrO_2 react with surface hydroxyl groups to produce highly reactive hydroxyl radicals ($*OH$) (Equation 7). These photogenerated holes can lead to dissociation of water molecules in the aqueous solution, producing radicals (Equation 8). The highly reactive hydroxyl radicals ($*OH$) and superoxide radicals ($*O_2^-$) can react with methyl orange dye adsorbed on ZrO_2 nanostructures and lead to its degradation as represented in Equations 9 and 10.

$$ZrO_2 + h\upsilon \rightarrow e^-(CB) + h^+(VB)$$

(5)

$$O_2 + e^- \rightarrow *O_2^-$$

(6)

$$h^+ + OH^- \rightarrow *OH$$

(7)

$$h^+ + H_2O \rightarrow H^+ + *OH^-$$

(8)

$$* \mathrm{OH}^- + \text{ Methyl orange } \rightarrow \text{ Degradation products} \tag{9}$$

$$* \mathrm{O}_2^- + \text{ Methyl orange } \rightarrow \text{ Degradation products} \tag{10}$$

The processes leading to photocatalytic degradation of methyl orange and the mechanism over the mesoporous ZrO_2 nanostructures were represented in Figure 9. It was reported that the enhanced photocatalytic activity of mesoporous structure to crystalline ZrO_2 nanomaterial was due to the light harvesting capability, prolong the life of the photoinduced electron-hole pairs, and facilitate the reactant accessibility to surface active sites [11]. In the photocatalytic degradation process, the increase in photocatalytic activity is associated with efficient separation of photogenerated electrons and holes. If a surface defect state is able to trap electrons or holes, recombination can be restricted. The presence of oxygen-deficient $ZrO_{0.35}$ impurity in the m-ZrO_2 sample could also be responsible for the oxygen vacancies and they are acting as electron acceptors to trap electrons and interstitial oxygen act as shallow trappers for holes, both of which prevent the recombination of photogenerated electrons and holes, thereby increasing the efficiency in the m-ZrO_2 sample. In addition, the m-ZrO_2 sample clearly possess the large pores, which can effectively facilitate both higher reactant accessibility to the surface active sites and more efficient multiple light scattering inside the pore channels [59].

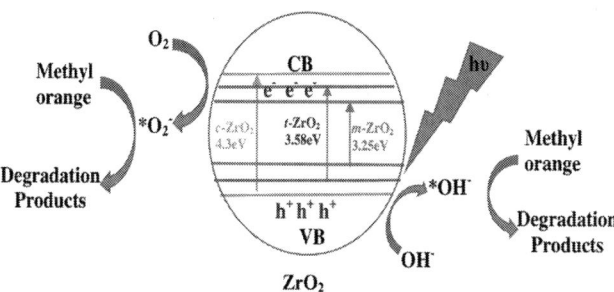

Figure 9: Schematic representation of the processes leading to photocatalytic degradation.

It is also known that the photocatalytic redox reaction mainly takes place on the surface of the photocatalysts and so the surface properties significantly influence the efficiency of catalyst [60]. Additionally, the surface hydroxyl groups of ZrO_2 are acidic to a certain degree, and the proportion of transformed azo structure increases into quinoid structure under acidic conditions (Figure 10). It is also reported that quinoid structure is more likely to be degraded than azo structure [61]. From the XRD patterns, it is clear that the m-ZrO_2 sample has the highest crystallinity (sharpest peaks, largest crystals). This is also reflected in the Raman spectra. A higher crystallinity is usually believed to be beneficial for photocatalysis because of the amount of defect sites in the structure (which usually act as recombination centers). In comparison, the m-ZrO_2 sample synthesized in this work possessed relatively high pore volume, pore size, and high density of hydroxyl groups.

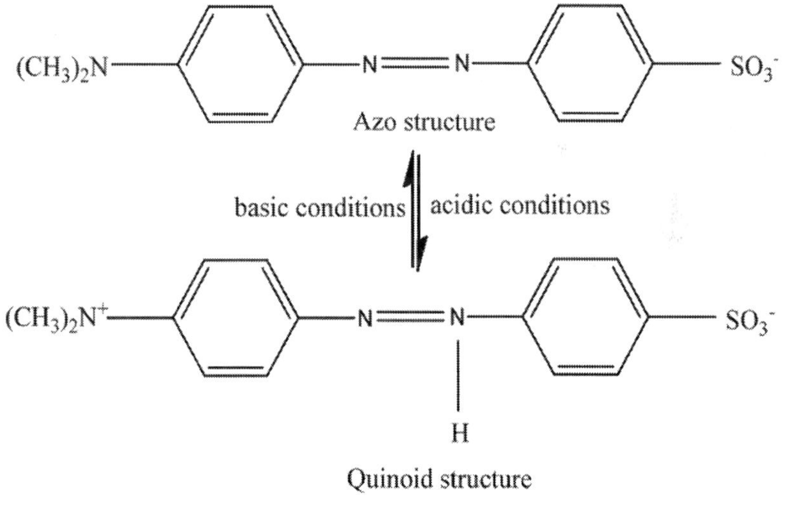

$(CH_3)_2N$———⬡———N＝N———⬡——— SO_3^-

Azo structure

basic conditions ⇅ acidic conditions

$(CH_3)_2N^+$———⬡———N＝N———⬡——— SO_3^-

H

Quinoid structure

Figure 10: Structural forms of methyl orange under acidic and basic conditions.

We also tested the reusability of the m-ZrO_2 catalyst for subsequent cycles of methyl orange degradation under optimized reaction conditions. Many of the reported photocatalysts have not been used for further degradation studies due to the fact that they undergo photocorrosion, by the direct illumination with light, and hence their photostability is diminished for further runs. For the reusability study, we collected the white-colored catalyst remained after the reaction, washed, dried at 100°C for 30 min, and used it for further reactions. The catalyst was found to be active for 5 cycles without any major deactivation, and more than 95% degradation was achieved in all experiments within 110 min using the m-ZrO_2 catalyst. The reusability of the m-ZrO_2 nanoparticles was ascribed to the low photocorrosive effect and high catalytic stability of the synthesized m-ZrO_2 sample.

CONCLUSIONS

Nanosize crystalline porous ZrO_2 nanoparticles with pure monoclinic, tetragonal, and cubic phases were synthesized by different preparation methods. The photocatalytic performance of the three ZrO_2 samples for the degradation of methyl orange was evaluated. Under the optimized reaction conditions, the m-ZrO_2 sample comparatively showed a higher methyl orange degradation activity than the t-ZrO_2 and c-ZrO_2 samples. The pronounced photocatalytic activity for m-ZrO_2 catalyst was mainly attributed to combining effects of factors including the presence of small amount of oxygen-deficient zirconium oxide phase, high crystallinity, broad pore size distribution, and high density of surface hydroxyl groups.

AUTHORS' CONTRIBUTIONS

All authors have contributed to the final manuscript of the present investigation. SB and KN have defined the research topic. TA, KN, and MM involved in the preparation, the characterization, and photocatalytic experiments. KN, TA, and MM wrote the manuscript. SB provided important suggestions on the draft of the manuscript. All authors examined and approved the final manuscript.

ACKNOWLEDGEMENTS

This project was funded by Saudi Basic Industries Corporation (SABIC) and the Deanship of Scientific Research (DSR), King Abdulaziz

University, Jeddah, under grant number MS/14-325-1433. The authors therefore acknowledge with thanks SABIC and DSR technical and financial support.

REFERENCES

1.	Roselin LS, Selvin R: Photocatalytic treatment and reusability of textile dyeing effluents from cotton dyeing industries.Sci Adv Mater. 2011, 3:113-9.

2.	Ameen S, Akhtar MS, Kim YS, Yang OB, Shin HS: Synthesis and characterization of novel poly(1-naphthylamine)/zinc oxide nanocomposites: application in catalytic degradation of methylene blue dye.Colloid Polym Sci. 2010, 288:1633-8.

3.	Neelakandeswari N, Sangami G, Dharmaraj N: Taek N K, Kim H Y: Spectroscopic investigations on the photodegradation of toluidine blue dye using cadmium sulphide nanoparticles prepared by a novel method.Spectrochim Acta, Part A 2011, 78:1592-1598.

4.	Hoffmann MR, Martin ST, Choi W, Bahnemann DW: Environmental applications of semiconductor photocatalysis.Chem Rev. 1995, 95:69-96.

5.	Kuriakose S, Satpati B, Mohapatra S: Enhanced photocatalytic activity of Co doped ZnO nanodisks and nanorods prepared by a facile wet chemical method.Phys Chem Chem Phys. 2014, 16:12741-9.

6.	Kuriakose S, Choudhary V, Satpati B, Mohapatra S: Facile synthesis of Ag-ZnO hybrid nanospindles for highly efficient photocatalytic degradation of methyl orange.Phys Chem Chem Phys 2014, 16:17560-17568.

7.	Cheng C, Amini A, Zhu C, Xu Z, Song H, Wang N: Enhanced photocatalytic performance of TiO_2-ZnO hybrid nanostructures. Sci Reports. 2014, 4:1481-6.

8.	Ren L, Li Y, Hou J, Zhao X, Pan C: Preparation and enhanced photocatalytic activity of TiO_2 nanocrystals with internal pores. ACS Appl Mater Interfaces. 2014, 6:1608-15.

9.	Navio JA, Hidalgo MC, Colon G, Botta SG, Litter MI: Preparation and physicochemical properties of ZrO_2 and Fe/ZrO_2 prepared by a sol-gel technique.Langmuir. 2001, 17:202-10.

10. Botta SG, Navio JA, Hidalgo MC, Restrepo GM, Litter MI: Photocatalytic properties of ZrO_2 and Fe/ZrO_2 semiconductors prepared by a sol-gel technique.J Photochem Photobiol A Chem. 1999, 129:89-99.

11. Sreethawong T, Ngamsinlapasathian S, Yoshikawa S: Synthesis of crystalline mesoporous-assembled ZrO_2 nanoparticles via a facile surfactant-aided sol-gel process and their photocatalytic dye degradation activity.Chem Eng J. 2013, 228:256-62.

12. Sohn JR, Ryu SG: Surface characterization of chromium oxide-zirconia catalyst.Langmuir. 1993, 9:126-31.

13. Gao PT, Meng LJ, dos Santos MP, Teixeira V, Andritschky M: Study of ZrO_2-Y_2O_3 films prepared by RF magnetron reactive sputtering. Thin Solid Films. 2000, 377:32-6.

14. Ma ZY, Yang C, Wei W, Li WH, Su YH: Surface properties and CO adsorption on zirconia polymorphs.J Mol Catal A Chem. 2005, 227:119-24.

15. Pokrovski K, Jung KT, Bell AT: Investigation of CO and CO_2 adsorption on tetragonal and monoclinic zirconia.Langmuir 2001, 17:4297-4303.

16. Nawale AB, Kanhe NS, Bhoraskar SV, Mathe VL, Das AK: Influence of crystalline phase and defects in the ZrO_2 nanoparticles synthesized by thermal plasma route on its photocatalytic properties.Mater Res Bull. 2012, 47:3432-9.

17. Zhao J, Wang X, Zhang L, Hou X, Li Y, Tang C: Degradation of methyl orange through synergistic effect of zirconia nanotubes and ultrasonic wave.J Hazar Mater. 2011, 188:231-4.

18. Ismail S, Ahmad ZA, Berenov A, Lockman Z: Effect of applied voltage and fluoride ion content on the formation of zirconia nanotube arrays by anodic oxidation of zirconium.Corros Sci. 2011, 53:1156-64.

19. Jiang W, He J, Zhong J, Lu J, Yuan S, Liang B: Preparation and photocatalytic performance of ZrO_2 nanotubes fabricated with anodization process.Appl Sur Sci. 2014, 307:407-13.

20. Shu Z, Jiao X, Chen D: Hydrothermal synthesis and selective photocatalytic properties of tetragonal star-like ZrO_2 nanostructures.Cryst Eng Comm. 2013, 15:4288-94.

21. Guo GY, Chen YL: A nearly pure monoclinic nanocrystalline zirconia.J Solid State Chem. 2005, 178:1675-82.

22. Rezaei M, Alavi SM, Sahebdelfar S, Xinmei L, Yan ZF: Synthesis of mesoporous nanocrystalline zirconia with tetragonal crystallite phase by using ethylene diamine as precipitation agent.J Mater Sci. 2007, 42:7086-92.

23. Tahir MN, Gorgishvili L, Li J: Facile synthesis and characterization of monocrystalline cubic ZrO_2 nanoparticles.Solid State Sci. 2007, 9:1105-9.

24. Calafat A: The influence of preparation conditions on the surface area and phase formation of zirconia.Stud Surf Sci Catal. 1998, 118:837-43.

25. Srinivasan R, De Angelis RJ, Ice G, Davis BH: Identification of tetragonal and cubic structures of zirconia using synchrotron x-radiation source.J Mater Res. 1991, 6:1287-92.

26. Abdala PM, Fantini MC, Craievich AF, Lamas DG: Crystallite size-dependent phases in nanocrystalline ZrO_2-Sc_2O_3.Phys Chem Chem Phys. 2010, 12:2822-9.

27. Mokhtar M, Basahel SN, Ali TT: Effect of synthesis methods for mesoporous zirconia on its structural and textural properties.J Mater Sci. 2013, 48:2705-13.

28. Bersani D, Lottici PP, Rangel G, Ramos E, Pecchi G, Gomez R, et al.: Micro-Raman study of indium doped zirconia obtained by sol-gel.J Non-Crystalline Solids. 2004, 345–346:116-9.

29. Gazzoli D, Mattei G, Valigi M: Raman and X-ray investigations of the incorporation of Ca^{2+} and Cd^{2+} in the ZrO_2 structure.J Raman Spectrosc. 2007, 38:824-31.

30. Chervin CN, Clapsaddle BJ, Chiu HW, Gash AE, Satcher JH, Kauzlarich SM: Aerogel synthesis of yttria-stabilized zirconia by a non-alkoxide sol-gel route.Chem Mater. 2005, 17:3345-51.

31. Kontoyannis CG, Orkoula M: Quantitative determination of the cubic, tetragonal and monoclinic phases in partially stabilized zirconias by Raman spectroscopy.J Mater Sci. 1994, 29:5316-20.

32. Ray JC, Patil RK, Pramanik P: Chemical synthesis and structural characterization of nanocrystalline powders of pure zirconia and yttria stabilized zirconia (YSZ).J Eur Ceram Soc. 2000, 20:1289-95.

33. Garvie RC. The occurrence of metastable tetragonal zirconia as a crystallite size effect. J Phys Chem. 1965;69:1238–43.

34. Zhao N, Pan D, Nie W, Ji X: Two-phase synthesis of shape-controlled colloidal zirconia nanocrystals and their characterization. J Am Chem Soc. 2006, 128:10118-24.

35. Kasatkin I, Girgsdies F, Ressler T, Caruso RA, Schattka JH, Urban J, et al.: HRTEM observation of the monoclinic-to-tetragonal (m-t) phase transition in nanocrystalline ZrO_2. J Mater Sci. 2004, 39:2151-7.

36. Inorganic Crystal Structure Database, FIZ Karlsruhe and the National Institute of Standards and Technology, Karlsruhe. 2014, http://icsd.fiz-karlsruhe.de/search/index.xhtml. Accessed 10 Sept 2014.

37. Rouquerol F, Rouquerol J, Sing K: Adsorption by powders and porous solid: principle, methodology, and applications. Academic, San Diego; 1999.

38. McBain JW: An explanation of hysteresis in the hydration and dehydration of gels. J Am Chem Soc. 1935, 57:699-700.

39. Basahel SN, Ali TT, Narasimharao K, Bagabas AA, Mokhtar M: Effect of iron oxide loading on the phase transformation and physicochemical properties of nanosized mesoporous ZrO_2. Mater Res Bull. 2012, 47:3463-72.

40. Wang W, Guo HT, Gao JP, Dong XH, Qin QX: XPS, UPS and ESR studies on the interfacial interaction in $Ni-ZrO_2$ composite plating. J Mater Sci. 2000, 35:1495-9.

41. Ardizzone S, Bianchi CL: XPS characterization of sulphated zirconia catalysts: the role of iron. Surf Inter Anal. 2000, 30:77-80.

42. Dongare MK, Dongare AM, Tare VB, Kemniz E: Synthesis and characterization of copper-stabilized zirconia as an anode material for SOFC. Solid State Ionics. 2002, 455:152-6.

43. Kawasaki KA: Positions of photoelectron and auger lines on the binding energy scale. XPS International, Japan; 1997.

44. Ram S, Mondal A: X-ray photoelectron spectroscopic studies of Al^{3+} stabilized $t-ZrO_2$ of nanoparticles. Appl Sur Sci. 2004, 221:237-47.

45. Gredelj S, Gerson AR, Kumar S, Cavallaro GP: Characterization of aluminium surfaces with and without plasma nitriding by X-ray photoelectron spectroscopy.Appl Surf Sci. 2001, 174:240-50.

46. Ardizzone S, Cattania MG, Lazzari P, Sarti M: Bulk, surface and double layer properties of zirconia polymorphs subjected to mechanical treatments.Mater Chem Phys. 1991, 28:399-412.

47. Rashad MM, Baioumy HM: Effect of thermal treatment on the crystal structure and morphology of zirconia nanopowders produced by three different routes.J Mater Process Tech. 2008, 195:178-85.

48. Herrera G, Montoya N, Domenech-Carbo A, Alarcon J: Synthesis, characterization and electrochemical properties of iron-zirconia solid solution nanoparticles prepared using a sol-gel technique. Phys Chem Chem Phys. 2013, 15:19312-21.

49. Li N, Dong B, Yuan W, Gao Y, Zheng L, Huang Y, et al.: ZrO_2 nanoparticles synthesized using ionic liquid microemulsion.J Dispersion Sci Technol. 2007, 28:1030-3.

50. French RH, Glass SJ, Ohuchi FS, Xu YN, Ching WY: Experimental and theoretical determination of the electronic structure and optical properties of three phases of ZrO_2.Phys Rev B Condens Matter. 1994, 49:5133-42.

51. Emeline V, Kuzmin GN, Purevdorj D, Ryabchuk VK, Serpone N: Spectral dependencies of the quantum yield of photochemical processes on the surface of wide band gap solids. 3. Gas/solid systems.J Phys Chem B 2000, 104:2989-99.

52. Chang SM, Doong RA: Inter band transitions in sol-gel-derived ZrO_2 films under different calcination conditions.Chem Mater. 2007, 19:4804-10.

53. Wang WW, Zhu YJ, Yang LX: ZnO-SnO_2 hollow spheres and hierarchical nanosheets: hydrothermal preparation, formation mechanism, and photocatalytic properties.Adv Funct Mater. 2007, 17:59-64.

54. Bachiller-Baeza B, Rodriguez-Ramos I, Guerrero-Ruiz A: Interaction of carbon dioxide with the surface of zirconia polymorphs.Langmuir. 1998, 14:3556-64.

55. Ma Z-Y, Yang C, Wei W, Li W-H, Sun Y-H: Surface properties and CO adsorption on zirconia polymorphs.J Mol Catal A Chem. 2005, 227:119-24.

56.   Guo B, Shen H, Shu K, Zeng Y, Ning W: The study of the relationship between pore structure and photocatalysis of mesoporous TiO_2.J Chem Sci. 2009, 121:317-21.

57.   Shao GN, Imran SM, Jeon SJ, Engole M, Abbas N, Haider MS, et al.: Sol-gel synthesis of photoactive zirconia-titania from metal salts and investigation of their photocatalytic properties in the photodegradation of methylene blue.Powder Technol. 2014, 258:99-109.

58.   Ahmed S, Rasul MG, Brown R, Hashi MA: Influence of parameters on the heterogeneous photocatalytic degradation of pesticides and phenolic contaminants in wastewater: a short review.J Environ Manage. 2011, 92:311-30.

59.   Jantawasu P, Sreethawong T, Chavadej S: Photocatalytic activity of nanocrystalline mesoporous-assembled TiO_2 photocatalyst for degradation of methyl orange monoazo dye in aqueous wastewater.Chem Eng J. 2009, 155:223-33.

60.   Xu C, Mei L: Synthesis and enhanced photocatalytic activity of hierarchical ZnO nanostructures.J Nanosci Nanotech. 2013, 13:513-6.

61.   Rhodes MD, Bell AT: The effects of zirconia morphology on methanol synthesis from CO and H_2 over Cu/ZrO_2 catalysts Part I.Steady-state studies. J Catal. 2005, 233:198-209.

Synthesis of Multi-shelled Zno Hollow Microspheres and Their Improved Photocatalytic Activity

Xiangyun Zeng[1, 2], Jiao Yang[1, 2], Liuxue Shi[1, 2], Linjie Li[1, 2], and Meizhen Gao[1, 2]

[1]Key Laboratory for Magnetism and Magnetic Materials of Ministry of Education, Lanzhou University, 730000 Lanzhou, People's Republic of China

[2]School of Physical Science and Technology, Lanzhou University, Lanzhou, Gansu 730000, People's Republic of China

ABSTRACT

Herein, we report an effective, facile, and low-cost route for preparing ZnO hollow microspheres with a controlled number of shells composed of small ZnO nanoparticles. The formation mechanism of multiple-shelled structures was investigated in detail. The number of shells is manipulated by using different diameters of carbonaceous microspheres. The products were characterized by X-ray powder diffraction, scanning electron microscopy, and transmission electron microscopy. The as-prepared ZnO hollow microspheres and ZnO nanoparticles were then used to study the degradation of methyl orange (MO) dye under ultraviolet (UV) light irradiation, and the triple-shelled ZnO hollow microspheres exhibit the best photocatalytic activity. This work is helpful to develop ZnO-based photocatalysts with high photocatalytic performance in addressing environmental protection issues, and it is also anticipated to other multiple-shelled metal oxide hollow microsphere structures.

BACKGROUND

With the sustainable development of industry and society, the contamination of the environment caused by organic pollutants is becoming an overwhelming problem all over the world [1]. In recent years, conventional biological, physical, and chemical treatment methods have been studied widely. Since the semiconductor-based photochemical electrode reported by Fujishima and Honda in 1972[2], photoactive nanomaterials as photocatalysts, especially semiconductor nanomaterials, have attracted the most attention to the degradation of organic compounds for the purpose of purifying wastewater. This is due to their high photocatalytic activity and excellent chemical and mechanical stability, and it is also an easy way to utilize the energy of solar light, abundantly available everywhere in the world [3]. Naturally, it is of substantial importance to carry out works related to semiconductor photocatalysis. Among these wide-bandgap semiconductors used in photoelectrochemical and photocatalytic applications, ZnO plays an important role in degrading various organic pollutants and photodegradation of bacteria due to its high catalytic activity, low cost, and environmental friendliness [4-6]. However, ZnO is a semiconductor with a bandgap of 3.37 eV and a large exciton binding energy of 60 meV at room temperature which results in the poor utilization of sunlight, limiting its photocatalytic efficiency. Therefore, it is essential to improve the photocatalytic properties of ZnO. According to the principle of photocatalysis, much research focuses on enhancing the surface areas of semiconductor nanomaterials by developing nanoscaled or porous appearance because a large surface area can achieve stronger light harvesting and provide more active sites at which the photocatalytic reaction occurs, small nanoparticles shorten the distance that electrons and holes migrate from bulk to reaction active sites to lower the possibility of recombination of the photogenerated charges, and the porosity can improve the photon application efficiency [7]. Thus, a number of efforts have been attracted to obtain high catalytic activity by manufacturing different ZnO nanostructures, such as nanoneedles [8], nanowires [9,10], nanorods [11,12], nanotetrapods [13], nanoplatelets [14], nanotubes [15,16], nanotowers [17], nanoflowers [18-20], and hollow nanospheres [21,22].

Hollow spheres and hollow spheres with multiple shells are of great interest in many current and emerging areas of technology because their unique structures enable physical properties such as uniform size, low density, and large surface area, which make them attractive materials for applications, such as sensors [23-26], catalysis [27-29], drug delivery [30-34], energy conversion [35-39], and storage systems [40-43]. For example, studies have demonstrated that multiple-shelled -Fe_2O_3 hollow spheres are much more sensitive than -Fe_2O_3 hollow spheres [44]. Multiple-shelled Co_3O_4 hollow microspheres were prepared and reported to have excellent cycle performance and enhanced lithium storage capacity [45]. Dong et al. reported the synthesis of quintuple-shelled SnO_2 hollow microspheres, and the quintuple-shelled SnO_2 hollow microspheres exhibited high-performance dye-sensitized solar cells [46]. Multiple-shelled metal oxides have been regarded as fascinating nanomaterials in the field of photocatalysis, partly due to their high quantum yield but also because the nanostructure gives a large surface-area-to-volume ratio and strong light-harvesting capabilities [47]. As a result, the study of the ZnO hollow nanosphere and hollow spheres with multiple shells of photocatalyst is both important and interesting.

Herein, we report a simple and general method to successfully fabricate ZnO hollow microspheres with a controlled number of shells by using carbonaceous saccharide microspheres with different diameters as templates. The photocatalytic property of the as-synthesized products is investigated by studying the degradation of methyl orange (MO) dye, and the triple-shelled ZnO hollow spheres with high surface area were proven to have excellent photocatalytic activity. The mechanism of formation of multiple-shelled ZnO hollow spheres and the reason for the high photocatalytic activity were also investigated.

METHODS

All chemicals were of reagent grade and were used as raw materials without further purification.

Synthesis

In this work, carbonaceous saccharide microspheres were used as sacrificial templates and zinc nitrate hexahydrate (Zn $(NO_3)_2 \cdot 6H_2O$) were used as metal precursors. Taking the synthesis of single-shelled ZnO hollow microspheres as an example, the typical synthesis is described as follows. Carbonaceous microspheres were synthesized through the emulsion polymerization reaction of sugar under hydrothermal conditions as described elsewhere [48,49]. The diameter of the obtained carbon spherules could be controlled through regulating the concentration of the sugar solution and reaction time. The carbonaceous saccharide microspheres were washed several times with absolute ethanol and deionized water until the filtrate was clear. Newly prepared carbonaceous microspheres (0.5 g) with diameters about 500 nm were dispersed in 1.5 M zinc nitrate solution (water/ethanol = 1:3, v/v, 25 mL) with the aid of ultrasonication. After ultrasonic dispersion for 0.5 h, the resulting suspension was aged for 8 h at 60°C in a water bath, vacuum filtered, washed with deionized water for several times, and then dried at 80°C in an oven for 12 h. In order to remove the templates, the resulting black composite microspheres were heated to 350°C for 1 h and then the temperature was raised to 450°C in air at the rate of 1°C min^{-1} and kept at 450°C for 2 h. The single-shelled ZnO hollow microspheres were prepared as a white powder product after the tube furnace cooled down to room temperature naturally. The synthesis processes of double- and triple-shelled ZnO hollow microspheres were also synthesized by following a similar procedure. The ZnO nanoparticles were synthesized by combustion method. In brief, 3 g Zn $(CH_3COO)_2$ $2H_2O$ and 1 g CO $(NH_2)_2$ were dissolved into 5 mL deionized water, and then the NH_3 H_2O was dropwise added into the solution until the solution turned into highly viscous gel precursors. Then the obtained viscous gel precursors were heated quickly at 500°C, and the precursors spontaneously ignited to produce white ZnO powders.

Characterization

The crystal phase, morphology, and composition of the produced products were characterized by X-ray powder diffraction (XRD; Rigaku RINT2400, Rigaku, Tokyo, Japan) with Cu K radiation ($s = 1.5418$ Å), field-emission scanning electron microscopy (FE-SEM; Hitachi S-4800, Hitachi, Tokyo, Japan), and transmission electron microscopy (TEM; FEI Tecnai G2 F30, FEI, Oregon, USA). The Fourier transform infrared (FTIR) spectrum of the precursor was recorded between 400 and 4,000 cm^{-1} on a Nicolet NEXUS 670 FTIR spectrometer (Thermo Fisher Scientific, Massachusetts, USA). Thermogravimetric analysis (TGA) was carried out in air at a heating rate of 1.00°C min^{-1} from 35.00°C to 700.00°C using a PerkinElmer Diamond TG/DTA instrument (PerkinElmer, Massachusetts, USA). Diffuse reflectance absorption spectra (DRS) were measured by a PerkinElmer 950 UV–vis spectrophotometer equipment (PerkinElmer, Massachusetts, USA), and the recorded range of the spectra is 250 to 800 nm.

Photoelectrochemical characterizations were carried out in a three-electrode system with the aid of the electrochemical workstation CHI 660 (CH Instruments, Texas, USA) and a conventional three-electrode system. Indium tin oxide covered by a thin film of the samples was used for the working electrode, with an active area of 1 cm^2. Platinum and Ag/AgCl (saturated KCl) electrode were used as the working electrode, auxiliary electrode, and reference electrode, respectively.

MO decomposition tests were carried out to study the photocatalytic activities of the as-synthesized products. Typically, 20 mg of photocatalysts was added into 300 mL of aqueous solution of the MO dyes (10 mg L^{-1}), and the solution was simultaneously sonicated and shaken for 10 min in an ultrasonic cleaning bath. Prior to the irradiation, the suspensions were magnetically stirred in the dark for 30 min to realize adsorption equilibrium; afterward, the photoreaction vessel was exposed to UV irradiation (500-W high-pressure Hg lamp with the main wavelength at 365 nm) under ambient conditions to start the photocatalytic reaction. At given time intervals, a certain volume of suspension (approximately 3 mL) was withdrawn. With the help of centrifugation, the catalyst was recovered and the MO concentration of the sample at different intervals was monitored using a spectrophotometer (WFJ-7200, Unico, Franksville, WI, USA). The cycling runs were carried out to demonstrate the stability and reusability of triple-shelled ZnO hollow microspheres. After one cycle, the photocatalyst was filtrated and washed drastically with deionized water, and then the succeeding cycling runs are the same as the first cycle.

RESULTS AND DISCUSSION

Figure 1a, b, c shows the SEM images of the monodispersed carbonaceous templates with three sizes prepared using the hydrothermal method at 200°C under different concentrations of sugar solution from 0.25 to 1 mol L^{-1}. A perfect sphere shape, a uniform diameter size, and monodispersed spherules can be seen. The diameter of the monodispersed carbon spherules varies from 0.5 to 4 μm. It indicates that the concentrations of the sugar solution play an important role in the diameter of the spherules. Figure 1d shows a TEM image of the fragment of carbon spherules. Clearly, there are large quantities of micropores of the carbonaceous templates. The key to the formation of the micropores is the escape of water through the flexible dewatering sugar in the experiment. Energy-dispersive X-ray spectroscopy (EDS) shows that the carbon spherules consisted of C and O elements. The strong absorption in the intensity of the bands at 3,433 cm^{-1} is the characteristic of the O-H stretching mode. The absorption bands lying at 1,703 and 1,625 cm^{-1} are associated with the stretching of the $C=O$ and $C=C$ groups, respectively. The absorption band at 1,000 to 1,300 cm^{-1} indicates the appearance of the stretching of C-OH. The FTIR spectra provide a clear evidence that the carbon spherules are rich with surface functional groups, which play an important role in the adsorption of the metal ion [19].

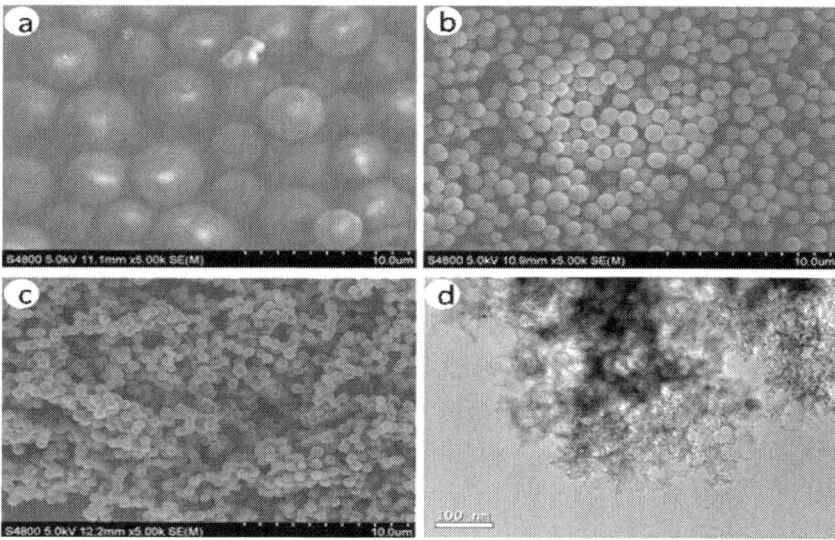

Figure 1: SEM images of monodispersed carbonaceous templates and TEM image of the as-prepared carbon spherules. (a-c) SEM images of monodispersed carbonaceous templates prepared under different glucose concentrations: (a) 1 mol L^{-1}, (b) 0.5 mol L^{-1}, (c) 0.25 mol L^{-1}.(d) TEM image of the as-prepared carbon spherules.

In addition, the precursor is characterized by TGA. there is a sharp mass loss from room temperature to 250°C in air, indicating that the start temperature of the decomposition of the precursor is around 250°C. The results of the TGA suggest that the carbon template could be completely removed. According to the TGA curve, we obtained the ZnO multi-shelled hollow spheres by choosing a temperature of 450°C for the thermal treatment of the precursor to ensure its complete decomposition.

The evolution process of shell formation was obtained by carrying out the reactions at the same temperatures for various times. The carbonaceous microspheres are full of zinc ions, heated to 450°C at the rate of 1°C min^{-1}. The TEM images of the products were obtained after different holding times in the heating process. After heating to 450°C for 0.5 h, the carbonaceous microspheres have no obvious change. When the holding time increases to 1 h, a shell around a solid sphere was formed. As testified by the XRD and EDS data, the resultant materials were obtained without carbon and had a triple-shelled hollow structure. Figure 2 also illustrates the general process of fabricating triple-shelled hollow zinc oxide microspheres. In order to increase the surface functional groups of carbonaceous microspheres, such as hydroxyl or carboxylic acid, which are important and necessary for the adsorption of zinc ions, the carbonaceous microspheres were immersed in 1 mol L^{-1} aqueous HCl for 24 h. Because of the differential shrinking rate of carbonaceous particles and zinc precursors, multiple-shelled ZnO hollow spheres were generated. The amount of metal ion adsorbed by the carbonaceous microspheres is the crucial factor, which determines the number of shells of hollow spheres. It is easy to understand that the carbonaceous microspheres are larger and much more zinc ions are absorbed inside their inner core, which lead to hollow spheres with more shells. Based on this, it is feasible to control the number of shells of ZnO hollow microspheres by using the carbonaceous microsphere temples with different diameters. The crystalline structures of the prepared ZnO hollow microspheres were investigated by XRD. From Figure 3, it can be seen that the XRD pattern of the obtained ZnO hollow microspheres has identical peaks, which is consistent with wurtzite-structured ZnO (JCPDS Card No. 36–1451) without any other phases present, indicating the formation of pure zinc oxide.

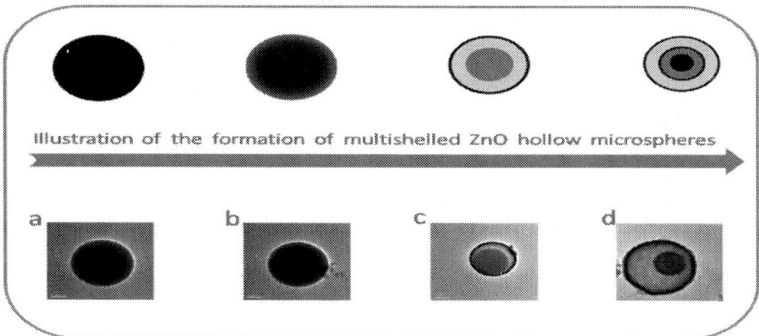

Figure 2: Schematic illustration and TEM images of the formation of multiple-shelled ZnO hollow microspheres. (a) 0 h. (b) 0.5 h. (c)1 h. (d) 2 h.

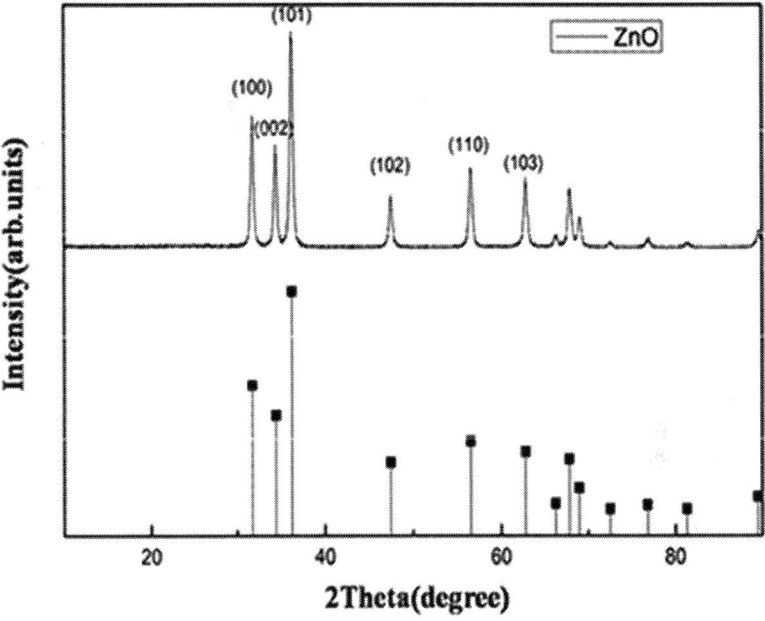

Figure 3: XRD pattern of the double-shelled ZnO hollow microspheres.

It is easy to discover the average values and standard deviations of the diameters of ZnO hollow spheres and nanoparticles. It can be seen that the diameter and number of shells of the as-prepared ZnO hollow spheres change when different carbonaceous saccharide microspheres are used. From the TEM and scanning TEM (STEM) (Figure 5h, i) images, it can be clearly seen that the spheres are with triple, double, and single shells. Some hollow spheres were broken, and the cross sections of the ZnO hollow spheres also show the number of ZnO hollow spheres (Figure 4e, f, g). The diameter of the ZnO hollow spheres is in the range of 0.5 to 4 μm and the number of shells in the range of 1 to 3. Figures 4d, h and 5d display SEM and TEM images of the as-prepared ZnO nanoparticles. Figure 5f, g shows a high-resolution TEM (HRTEM) image recorded from the outer shell of one hollow sphere and single-shelled hollow sphere. It can be observed that the shell is porous and composed of crystalline nanoparticles. Figure 5j shows the HRTEM image of triple-shelled hollow nanospheres, from which we determined the interplanar distances to be 0.248 nm, which corresponds well to the lattice spacing of the (101) plane of wurtzite

ZnO. Figure 5k reveals that these nanostructures are mainly composed of the elements Zn and O, indicating that pure ZnO nanospheres were obtained, and the elemental mapping of ZnO hollow spheres also shows the composition and element distribution of the ZnO hollow spheres. The result is in accord with the XRD data.

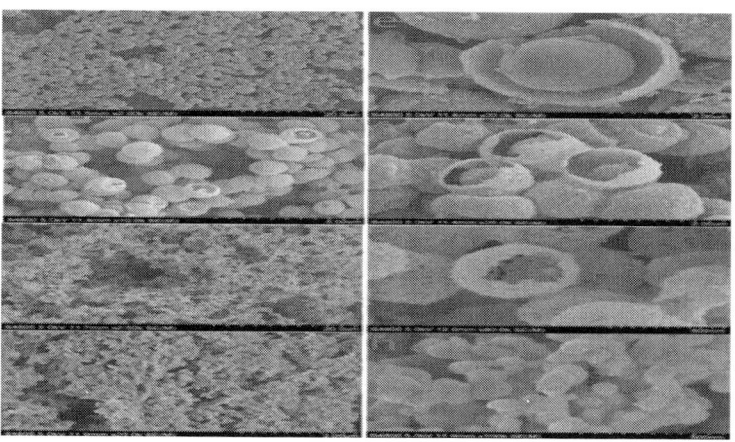

Figure 4: Low- and high-magnification SEM images of the triple-, double-, and single-shelled ZnO hollow spheres and ZnO nanoparticles. (a, e) Triple-, (b, f) double-, (c, g) and single-shelled ZnO hollow spheres and (d, h) ZnO nanoparticles.

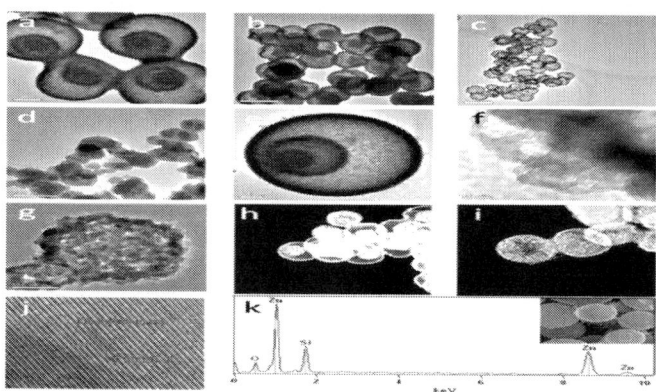

Figure 5: TEM, HRTEM, and STEM images and EDS spectrum of ZnO hollow spheres. (a-d) TEM images of the triple-, double-, and single-shelled ZnO hol-

low spheres and ZnO nanoparticles. (e-g) High-magnification TEM images of the triple- and single-shelled ZnO hollow spheres. (h, i) STEM images of the triple-, double-, and single-shelled ZnO hollow spheres. (j, k) HRTEM image and EDS spectrum of an individual triple-shelled ZnO hollow microsphere. The inset of Figure 5k is the SEM image of the sample, and the place marked by a rectangle shows the specific position of EDS analysis.

In order to testify the prediction that multiple-shelled hollow microsphere architectures can provide even more efficient multi-reflections of UV light within their interiors compared with nanoparticles and single-shelled and double-shelled hollow microspheres, the photocatalytic activities of the triple-shelled, double-shelled, and single-shelled ZnO hollow spheres and ZnO nanoparticles were measured and the photocatalytic decomposition of MO was carried out in a vessel containing a suspension of 20 mg in 500 mL of MO solution (10 mgL^{-1}) under UV light irradiation (500 W of the UV lamp). Before the irradiation, the suspensions were magnetically stirred in the dark for 30 min to ensure adsorption/desorption equilibrium. Photographs of MO under UV light irradiation for different periods of time are shown in Figure 6a. The triple-shelled ZnO hollow spheres behaved as effective catalysts in the photodegradation of MO (Figure 6b), and it can be seen that the MO solution was almost degraded completely within 30 min. Figure 6c shows the degradation rate of MO over different photocatalysts; ZnO nanoparticles (as shown in Figure 5d, e, f, g, h) exhibit the lowest activity, while triple-shelled ZnO hollow microspheres (as shown in Figure 5a, b, c, d, e) show the highest activity after irradiation for the same time. As expected, the triple-shelled ZnO hollow microspheres exhibit the best photodegradation behavior. Since all the samples possess similar crystalline structures, such an improvement of photocatalytic activity is governed by the unique multiple-shelled hollow morphology.

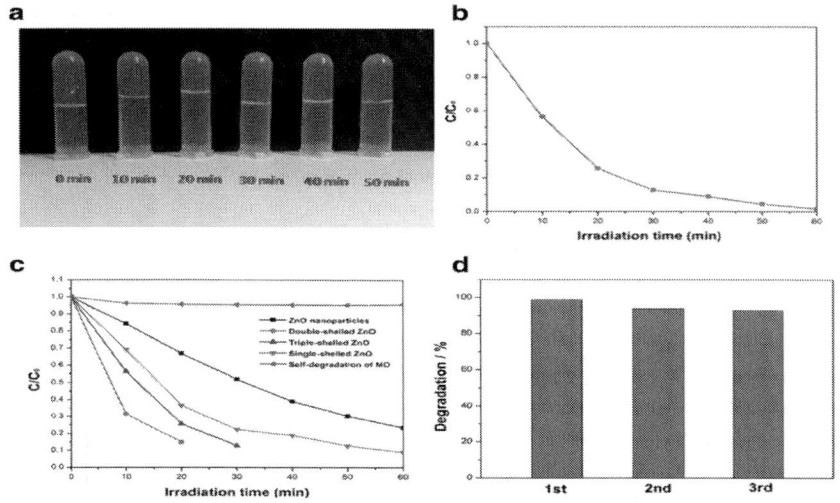

Figure 6: Photographs, concentrations, and photocatalytic degradation of MO. (a) Photographs of MO irradiated by UV light for different periods of time. (b) Concentration of MO in the solution with the triple-shelled ZnO hollow spheres. (c) Concentration of MO in the solution with different photocatalysts versus the 30-min exposure time to UV irradiation. (d) Cycling runs in the photocatalytic degradation of MO in the presence of the triple-shelled ZnO hollow spheres.

Figure 7 shows the room-temperature UV–vis DRS of four samples. The bandgap energies (Eg) calculated on the basis of the corresponding absorption edges are 3.14 eV (ZnO nanoparticles) and 3.08 eV (triple-shelled ZnO). Compared with ZnO nanoparticles, the ZnO hollow microspheres show redshifts of the absorption edge, and it is found that the wavelength for absorption edge increases with increasing number of the shell of ZnO hollow microspheres. The results may come from the raising of the base absorption line which mainly contributed to the diameter of the nanoparticles of ZnO hollow microspheres being less than the diameter of ZnO nanoparticles [7]. All the absorption edge of these samples is still in the UV light region. Therefore, the photocatalytic activities of the samples were measured using the Hg lamp with the main wavelength at 365 nm as the irradiation source. In order to further demonstrate this structure–property relationship, we fabricated film photodetector devices composed of the ZnO nanoparticles and the triple-, double-, and single-shelled ZnO hollow spheres and measured their reproducible responses to on-off light cycles, and the results are shown in Figure 8. It demonstrates that the photodetector devices composed of the triple-shelled ZnO hollow microspheres have a very fast response time and excellent repeatability. It was found that the photocurrent on the triple-shelled ZnO hollow microspheres is about 6.5 times higher than that on the ZnO nanoparticles. It indicated that the multiple-shelled ZnO hollow microspheres can be used in UV light photodetectors.According to the above experimental results and analysis, the reasons for the optimized photocatalytic activity and enhancement of the photocurrent for triple-shelled ZnO hollow microspheres were discussed in detail and the photocatalytic mechanism was proposed. The mechanism for degradation of MO caused by the triple-shelled ZnO hollow microspheres under UV light irradiation is shown in Figure 9. This can be ascribed to the following. Firstly, the triple-shelled hollow microsphere structures can enable multiple light reflection and scattering between the outer spherical shell and the two interior shells compared with the nanoparticles and single-shelled hollow spheres to provide a more efficient way to enhance light-harvesting efficiency. (The insets show a schematic illustration of Figure 6c). Secondly, this structure also can supply more specific surface area for adsorbing more dye molecules which contribute to the high degradation performance of the triple-shelled hollow spheres. Furthermore, the better photo-induced charge separation efficiency of triple-shelled hollow microsphere structures plays an important role in a significant enhancement in conductivity. Therefore, the triple-shelled ZnO hollow microsphere structures which are in favor of these features have much better photocatalytic activity than the ZnO nanoparticles and ZnO hollow spheres.

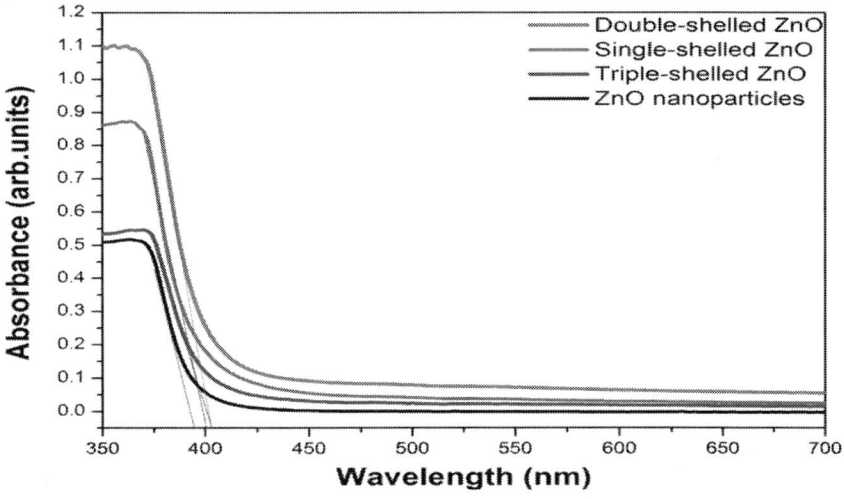

Figure 7: The room-temperature UV–vis DRS of ZnO nanoparticles and the triple-, double-, and single-shelled ZnO hollow spheres.

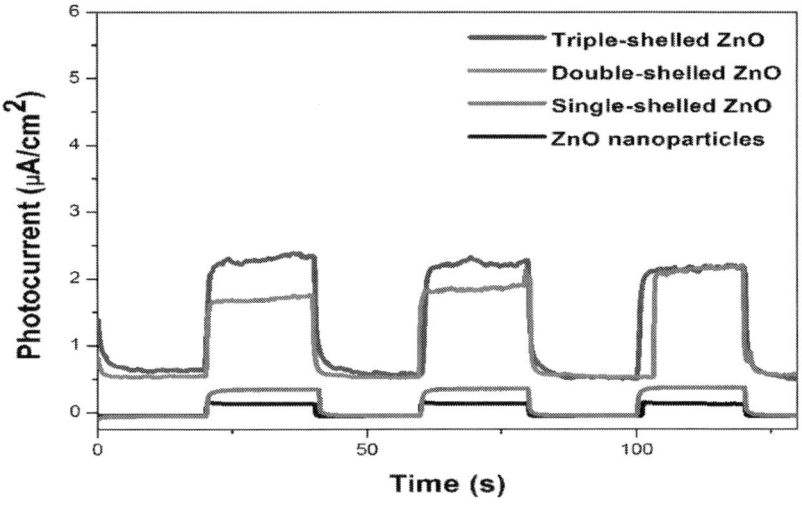

Figure 8: Photocurrent response of the ZnO nanoparticles and the triple-, double-, and single-shelled ZnO hollow spheres.

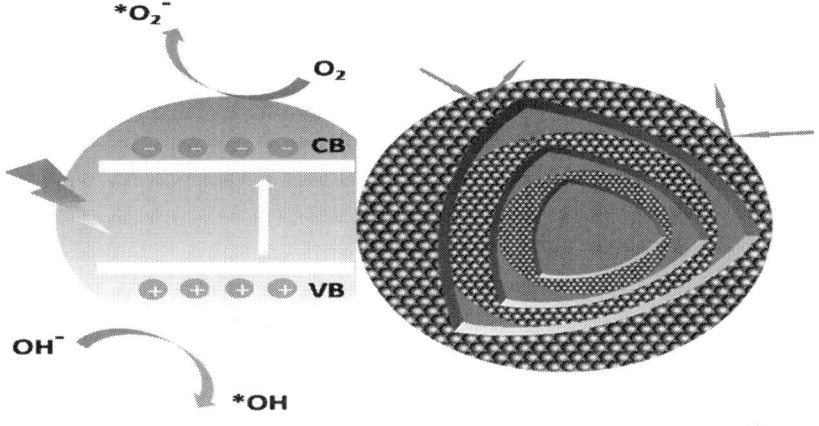

Figure 9: Schematic diagrams showing the photocatalytic performance of the triple-shelled ZnO hollow spheres.

CONCLUSIONS

In summary, it has been demonstrated that different numbers of shells of ZnO hollow microspheres can be successfully prepared by a facile route. Compared with nanoparticles and single-shelled hollow spheres, the obviously improved photocurrent responses and enhanced photocatalytic activity of the triple-shelled hollow microspheres have been achieved. The reason for this is attributed to the multiple-shelled hollow microsphere structure that captures large numbers of ultraviolet light photons and has large surface areas for absorbing more dye molecules. The multiple-shelled ZnO hollow nanospheres can be regarded as an excellent photocatalyst candidate and may be used in other fields such as UV and gas sensors and solar cells. This new synthetic concept is also helpful to controllably construct other multiple-shelled metal oxide hollow microsphere structures with enhanced properties for microelectronics, optoelectronics, and other applications.

AUTHORS' CONTRIBUTIONS

The experiments and characterization presented in this work were carried out by XZ, JY, LS, and LL. The experiments were designed by XZ and MG. XZ, JY, and MG analyzed and discussed the results obtained from the experiments. The manuscript was prepared by XZ. MG helped with the draft editing. All authors read and approved the final manuscript.

ACKNOWLEDGEMENTS

This work was financially supported by the National Natural Science Foundation of China (51371093) and Program for Changjiang Scholars and Innovative Research Team in University (IRT1251).

REFERENCES

1. Elliott JE, Elliott KH: Tracking marine pollution. *Science* 2013, 340:556.

2. Fujishima A: Electrochemical photolysis of water at a semiconductor electrode. *Nature* 1972, 238:37-38.

3. Hoffmann MR, Martin ST, Choi W, Bahnemann DW: Environmental applications of semiconductor photocatalysis. *Chem Rev* 1995, 95:69-96.

4. Karunakaran C, Rajeswari V, Gomathisankar P: Enhanced photocatalytic and antibacterial activities of sol–gel synthesized ZnO and Ag-ZnO. *Mater Sci Semicond Process* 2011, 14:133-138.

5. Akhavan O, Azimirad R, Safa S: Functionalized carbon nanotubes in ZnO thin films for photoinactivation of bacteria. *Mater Chem Phys* 2011, 130:598-602.

6. Wahab R, Kim Y-S, Mishra A, Yun S-I, Shin H-S: Formation of ZnO micro-flowers prepared via solution process and their antibacterial activity. *Nanoscale Res Lett* 2010, 5:1675-1681

7. Zhu C, Lu B, Su Q, Xie E, Lan W: A simple method for the preparation of hollow ZnO nanospheres for use as a high performance photocatalyst. *Nanoscale* 2012, 4:3060-3064.

8. Yang JL, An SJ, Park WI, Yi GC, Choi W: Photocatalysis using ZnO thin films and nanoneedles grown by metal–organic chemical vapor deposition. *Adv Mater* 2004, 16:1661-1664

9. Kuo T-J, Lin C-N, Kuo C-L, Huang MH: Growth of ultralong ZnO nanowires on silicon substrates by vapor transport and their use as recyclable photocatalysts. *Chem Mater* 2007, 19:5143-5147.

10. Jung S, Yong K: Fabrication of CuO–ZnO nanowires on a stainless steel mesh for highly efficient photocatalytic applications. *Chem Commun* 2011, 47:2643-2645.

11. Akhavan O: Graphene nanomesh by ZnO nanorod photocatalysts. *ACS Nano* 2010, 4:4174-4180

12. Lu X, Wang G, Xie S, Shi J, Li W, Tong Y, Li Y: Efficient photocatalytic hydrogen evolution over hydrogenated ZnO nanorod arrays. *Chem Commun* 2012, 48:7717-7719

13. Wan Q, Wang T, Zhao J: Enhanced photocatalytic activity of ZnO nanotetrapods. *Appl Phys Lett* 2005, 87:083105.

14. Ye C, Bando Y, Shen G, Golberg D: Thickness-dependent photocatalytic performance of ZnO nanoplatelets. *J Phys Chem B* 2006, 110:15146-15151.

15. Wang H, Li G, Jia L, Wang G, Tang C: Controllable preferential-etching synthesis and photocatalytic activity of porous ZnO nanotubes. *J Phys Chem C* 2008, 112:11738-11743.

16. Chu D, Masuda Y, Ohji T, Kato K: Formation and photocatalytic application of ZnO nanotubes using aqueous solution. *Langmuir* 2009, 26:2811-2815.

17. Deng D, Martin ST, Ramanathan S: Synthesis and characterization of one-dimensional flat ZnO nanotower arrays as high-efficiency adsorbents for the photocatalytic remediation of water pollutants. *Nanoscale* 2010, 2:2685-2691.

18. Barka-Bouaifel F, Sieber B, Bezzi N, Benner J, Roussel P, Boussekey L, Szunerits S, Boukherroub R: Synthesis and photocatalytic activity of iodine-doped ZnO nanoflowers. *J Mater Chem* 2011, 21:10982-10989.

19. Wang Y, Li X, Wang N, Quan X, Chen Y: Controllable synthesis of ZnO nanoflowers and their morphology-dependent photocatalytic activities. *Sep Purif Technol* 2008, 62:727-732

20. Lai Y, Meng M, Yu Y, Wang X, Ding T: Photoluminescence and photocatalysis of the flower-like nano-ZnO photocatalysts prepared by a facile hydrothermal method with or without ultrasonic assistance. *Appl Catal Environ* 2011, 105:335-345

21. Deng Z, Chen M, Gu G, Wu L: A facile method to fabricate ZnO hollow spheres and their photocatalytic property. *J Phys Chem B* 2008, 112:16-22.

22. Yu J, Yu X: Hydrothermal synthesis and photocatalytic activity of zinc oxide hollow spheres. *Environ Sci Technol* 2008, 42:4902-4907.

23. Li Z, Lai X, Wang H, Mao D, Xing C, Wang D: General synthesis of homogeneous hollow core – shell ferrite microspheres. *J Phys Chem C* 2009, 113:2792-2797.

24. Wang L, Lou Z, Fei T, Zhang T: Zinc oxide core–shell hollow microspheres with multi-shelled architecture for gas sensor applications. *J Mater Chem* 2011, 21:19331-19336

25. Hong YJ, Yoon JW, Lee JH, Kang YC: One–pot synthesis of Pd–loaded SnO_2 yolk–shell nanostructures for ultraselective methyl benzene sensors. *Chemistry* 2014, 20:2737-2741.

26. Yoon J-W, Hong YJ, Kang YC, Lee J-H: High performance chemiresistive H_2S sensors using Ag-loaded SnO_2 yolk–shell nanostructures. *RSC Advs* 2014, 4:16067-16074.

27. Xi G, Yan Y, Ma Q, Li J, Yang H, Lu X, Wang C: Synthesis of multiple–shell WO_3 hollow spheres by a binary carbonaceous template route and their applications in visible–light photocatalysis. *Chemistry* 2012, 18:13949-13953.

28. Zhang H, Du G, Lu W, Cheng L, Zhu X, Jiao Z: Porous TiO_2 hollow nanospheres: synthesis, characterization and enhanced photocatalytic properties. *Cryst Eng Comm* 2012, 14:3793-3801.

29. Song X, Gao L: Fabrication of hollow hybrid microspheres coated with silica/titania via sol–gel process and enhanced photocatalytic activities. *J Phys Chem C* 2007, 111:8180-8187.

30. Yang J, Lee J, Kang J, Lee K, Suh J-S, Yoon H-G, Huh Y-M, Haam S: Hollow silica nanocontainers as drug delivery vehicles. *Langmuir* 2008, 24:3417-3421.

31. Cao S-W, Zhu Y-J, Ma M-Y, Li L, Zhang L: Hierarchically nanostructured magnetic hollow spheres of Fe_3O_4 and $-Fe_2O_3$: preparation and potential application in drug delivery. *J Phys Chem C* 2008, 112:1851-1856.

32. Zhu Y, Fang Y, Kaskel S: Folate-conjugated Fe_3O_4 @ SiO_2 hollow mesoporous spheres for targeted anticancer drug delivery. *J Phys Chem C* 2010, 114:16382-16388.

33. Kawashima Y, Niwa T, Takeuchi H, Hino T, Itoh Y: Hollow microspheres for use as a floating controlled drug delivery system in the stomach. *J Pharm Sci* 1992, 81:135-140.

34. Sato Y, Kawashima Y, Takeuchi H, Yamamoto H: In vitro evaluation of floating and drug releasing behaviors of hollow microspheres (microballoons) prepared by the emulsion solvent diffusion method. *Eur J Pharm Biopharm* 2004, 57:235-243.

35. Qian J, Liu P, Xiao Y, Jiang Y, Cao Y, Ai X, Yang H: TiO_2–coated multilayered SnO_2 hollow microspheres for dye–sensitized solar cells. *Adv Mater* 2009, 21:3663-3667.

36. Wu X, Lu GQM, Wang L: Shell-in-shell TiO_2 hollow spheres synthesized by one-pot hydrothermal method for dye-sensitized solar cell application. *Energy Environ Sci* 2011, 4:3565-3572.

37. Du J, Qi J, Wang D, Tang Z: Facile synthesis of Au@ TiO_2 core–shell hollow spheres for dye-sensitized solar cells with remarkably improved efficiency. *Energy Environ Sci* 2012, 5:6914-6918.

38. XianáGuo C, MingáLi C: Nanoparticle self-assembled hollow TiO_2 spheres with well matching visible light scattering for high performance dye-sensitized solar cells. *Chem Commun* 2012, 48:8832-8834.

39. Wu D, Zhu F, Li J, Dong H, Li Q, Jiang K, Xu D: Monodisperse TiO_2 hierarchical hollow spheres assembled by nanospindles for dye-sensitized solar cells. *J Mater Chem* 2012, 22:11665-11671.

40. XianáGuo C, MingáLi C: Template-free bottom-up synthesis of yolk–shell vanadium oxide as high performance cathode for lithium ion batteries. *Chem Commun* 2013, 49:1536-1538

41. Zhang G, Yu L, Wu HB, Hoster HE, Lou XWD: Formation of $ZnMn_2O_4$ ball–in–ball hollow microspheres as a high–performance anode for lithium–Ion batteries. *Adv Mater* 2012, 24:4609-4613.

42. Cao AM, Hu JS, Liang HP, Wan LJ: Self–assembled vanadium pentoxide (V_2O_5) hollow microspheres from nanorods and their application in lithium–Ion batteries. *Angew Chem Int Ed* 2005, 44:4391-4395

43. Zhou L, Wu HB, Zhu T, Lou XWD: Facile preparation of $ZnMn_2O_4$ hollow microspheres as high-capacity anodes for lithium-ion batteries. *J Mater Chem* 2012, 22:827-829.

44. Lai X, Li J, Korgel BA, Dong Z, Li Z, Su F, Du J, Wang D: General synthesis and gas–sensing properties of multiple–shell metal oxide hollow microspheres. *Angew Chem Int Ed Engl* 2011, 123:2790-2793.

45. Wang J, Yang N, Tang H, Dong Z, Jin Q, Yang M, Kisailus D, Zhao H, Tang Z, Wang D:Accurate control of multishelled Co_3O_4 hollow microspheres as high–performance anode materials in lithium–ion batteries. *Angew Chem Int Ed Engl* 2013, 125:6545-6548.

46. Dong Z, Ren H, Hessel CM, Wang J, Yu R, Jin Q, Yang M, Hu Z, Chen Y, Tang Z: Quintuple–shelled SnO_2 hollow microspheres with superior light scattering for high–performance dye-sensitized solar cells. *Adv Mater* 2014, 26:905-909.

47. Lai X, Halpert JE, Wang D: Recent advances in micro-/n-structured hollow spheres for energy applications: from simple to complex systems. *Energy Environ Sci* 2012, 5:5604-5618.

48. Wang Q, Li H, Chen L, Huang X: Monodispersed hard carbon spherules with uniform nanopores. *Carbon* 2001, 39:2211-2214.

49. Sun X, Li Y: Colloidal carbon spheres and their core/shell structures with noble–metal nanoparticles. *Angew Chem Int Ed* 2004, 43:597-601.

Photocatalytic Reduction Synthesis of SrTiO3-graphene Nanocomposites and Their Enhanced Photocatalytic Activity

Tao Xian[1, 2], Hua Yang[1, 2], Lijing Di[2], Jinyuan Ma[1], Haimin Zhang[2], and Jianfeng Dai[1, 2]

[1]State Key Laboratory of Advanced Processing and Recycling of Nonferrous Metals, Lanzhou University of Technology, Lanzhou 730050, People's Republic of China

[2]School of Science, Lanzhou University of Technology, Lanzhou 730050, People's Republic of China

ABSTRACT

SrTiO$_3$-graphene nanocomposites were prepared via photocatalytic reduction of graphene oxide by UV light-irradiated SrTiO$_3$ nanoparticles. Fourier transformed infrared spectroscopy analysis indicates that graphene oxide is reduced into graphene. Transmission electron microscope observation shows that SrTiO$_3$ nanoparticles are well assembled onto graphene sheets. The photocatalytic activity of as-prepared SrTiO$_3$-graphene composites was evaluated by the degradation of acid orange 7 (AO7) under a 254-nm UV irradiation, revealing that the composites exhibit significantly enhanced photocatalytic activity compared to the bare SrTiO$_3$ nanoparticles. This can be explained by the fact that photogenerated electrons are captured by graphene, leading to an increased separation and availability of electrons and holes for the photocatalytic reaction. Hydroxyl ($\cdot$OH) radicals were detected by the photoluminescence technique using terephthalic acid as a probe molecule and were found to be produced over the irradiated SrTiO$_3$ nanoparticles and SrTiO$_3$-graphene composites; especially, an enhanced yield is observed for the latter. The influence of ethanol, KI, and N$_2$ on the photocatalytic efficiency was also investigated. Based on the experimental results, $\cdot$OH, h$^+$, and H$_2$O$_2$ are suggested to be the main active species in the photocatalytic degradation of AO7 by SrTiO$_3$-graphene composites.

BACKGROUND

Semiconductor photocatalysts have attracted considerable attention over the past decades due to their potential applications in solar energy conversion and environmental purification [1,2]. Among them, SrTiO$_3$, a well-known cubic perovskite-type multimetallic oxide with a bandgap energy (E_g) of approximately 3.2 eV, is proved to be a promising photocatalyst for water splitting and degradation of organic pollutants [3-6]. Furthermore, the photocatalytic activity of SrTiO$_3$ can be tailored or enhanced by doping with metalloid elements, decoration with noble metals, and composite with other semiconductors [7-10]. It is generally accepted that the basic principle of semiconductor photocatalysis involves the photogeneration of electron–hole (e$^-$-h$^+$) pairs, migration of the photogenerated carriers to the photocatalyst surface, redox reaction of the carriers with other chemical species to produce active species (such as $\cdot$OH, $\cdot O_2$, and H_2O_2), and attack of the active species on pollutants leading to their degradation. In these processes, the high recombination rate of the photogenerated carries greatly limits the photocatalytic activity of catalysts. Therefore, the effective separation of photogenerated electron–hole pairs is very important in improving the photocatalytic efficiency.

Graphene, being a two-dimensional (2D) sheet of sp^2-hybridized carbon atoms, possesses unique properties including high electrical conductivity, electron mobility, thermal conductivity, mechanical strength, and chemical stability [11-13]. On account of its outstanding properties, graphene has been frequently used as an ideal support to integrate with a large number of functional nanomaterials to form nanocomposites with improved performances in the fields of photocatalysts [14-21], supercapacitors [22], field-emission emitters [23], and fuel cells [24]. Particularly, the combination of graphene with photocatalysts is demonstrated to be an efficient way to promote the separation of photogenerated electron–hole pairs and then enhance their photocatalytic activity [14-21]. In these photocatalyst-graphene composites, photogenerated electrons can be readily captured by graphene which acts as an electron acceptor, leading to an increasing availability of photogenerated electrons and holes participating in the photocatalytic reactions. But so far, the investigation concerning the photocatalytic performance of $SrTiO_3$-graphene nanocomposites has been rarely reported.

Up to now, semiconductor-graphene nanocomposites have been generally prepared using graphene oxide as the precursor, followed by its reduction to graphene. To reduce the graphene oxide, several methods have been employed including chemical reduction using hydrazine or $NaBH_4$[14], high-temperature annealing reduction [15], hydrothermal reduction using supercritical water [16], green chemistry method [17], and photocatalytic reduction using semiconductors [18-21]. Among them, the photocatalytic reduction is an environment-friendly and a mild way for the synthesis of semiconductor-graphene composites. In this route, the solution containing the photocatalyst and graphene oxide is irradiated with light energy greater than the E_g of the photocatalyst, during which graphene oxide receives electrons from the excited photocatalyst and is thus reduced to graphene. During the photocatalytic reduction process, photocatalyst nanoparticles are assembled onto graphene sheets to form photocatalyst-graphene composites. Herein, we report the synthesis of $SrTiO_3$-graphene nanocomposites via the photocatalytic reduction method. The photocatalytic activity of the composites was evaluated by the degradation of acid orange 7 (AO7) under ultraviolet (UV) light irradiation, and the photocatalytic mechanism involved was discussed.

METHODS

$SrTiO_3$ nanoparticles were synthesized via a polyacrylamide gel route as described in the literature[25]. The graphene oxide used in this research was purchased from Nan-Jing XF Nano Materials Tech Co. Ltd. (Nanjing, China). $SrTiO_3$-graphene composites were prepared via a photocatalytic reduction route. A certain amount of graphene oxide was dispersed in 50 mL distilled water, followed by ultrasonic treatment of the suspension for 30 min. Then, 0.1 g $SrTiO_3$ nanoparticles and 0.0125 g ammonium oxalate (AO) were added to the suspension under magnetic stirring. After stirring for 10 min, the mixture was purged with nitrogen and exposed to UV light irradiation from a 15-W low-pressure mercury lamp for 5 h under mild stirring. During the irradiation, the color of the mixture changed from brown to black, indicating the reduction of the graphene oxide. After that, the product was separated from the reaction solution by centrifugation at 4,000 rpm for 10 min, washed several times with distilled water and absolute ethanol, and then dried in a thermostat drying oven at 60°C for 4 h to obtain $SrTiO_3$-graphene composites. A series of samples were prepared by varying the weight fraction of graphene oxide from 2.5% to 10%.

The photocatalytic activity of the samples was evaluated by the degradation of AO7 under UV light irradiation of a 15-W low-pressure mercury lamp (= 254 nm). The initial AO7 concentration was 5 mg L^{-1} with a photocatalyst loading of 0.5 g L^{-1}. Prior to irradiation, the mixed solution was ultrasonically treated in the dark to make the photocatalyst uniformly dispersed. The concentration of AO7 after the photocatalytic degradation was determined by measuring the absorbance of the solution at a fixed wavelength of 484 nm. Before the absorbance measurements, the reaction solution was centrifuged for 10 min at 4,000 rpm to remove the photocatalyst. The degradation percentage is defined as $(C_0 - C_t) / C_0 \times 100\%$, where C_0 and C_t are the AO7 concentrations before and after irradiation, respectively. To investigate the photocatalytic stability of the $SrTiO_3$-graphene composites, the recycling tests for the degradation of AO7 using the composite were carried out. After the first cycle, the photocatalyst was collected by centrifugation, washed with water, and dried. The recovered photocatalyst was introduced to the fresh AO7 solution for the next cycle of the photocatalysis experiment under the same conditions. The process was repeated four times.

Terephthalic acid (TA) was used as a probe molecule to examine hydroxyl ($\cdot$OH) radicals produced over the irradiated $SrTiO_3$-graphene composites. It is expected that TA reacts with $\cdot$OH to generate a highly fluorescent compound, 2-hydroxyterephthalic acid (TAOH). By measuring the photoluminescence (PL) intensity of TAOH that is pronounced around 429 nm, the information about $\cdot$OH can be obtained. TA was dissolved in a NaOH solution (1.0 mmol L^{-1}) to make a 0.25-mmol L^{-1} TA solution and then to the solution was added 0.5 g L^{-1} $SrTiO_3$-graphene composites. The mixed solution, after several minutes of ultrasound treatment in the dark, was illuminated under a 15-W low-pressure mercury lamp. The reacted solution was centrifuged for 10 min at 4,000 rpm to remove the photocatalyst and was then used for the PL measurements through a fluorescence spectrophotometer with the excitation wavelength of 315 nm.

The phase purity of the samples was examined by means of X-ray powder diffraction (XRD) with Cu K radiation. Fourier transform infrared spectroscopy (FTIR) measurements were carried out on a Bruker IFS 66v/S spectrometer (Ettlingen, Germany). The morphology of the samples was observed by a field emission transmission electron microscope (TEM). The UV-visible diffuse reflectance spectra were measured using a UV-visible spectrophotometer with an integrating sphere attachment.

RESULTS AND DISCUSSION

Figure 1 schematically shows the photocatalytic reduction process of graphene oxide by UV light-irradiated $SrTiO_3$ nanoparticles. It is noted that the $SrTiO_3$ particles have an isoelectric point at pH 8.5[26]; that is, they bear a negative surface charge when pH > 8.5 and a positive surface charge when pH < 8.5. When the $SrTiO_3$ particles are added to the graphene oxide suspension, the pH value of the mixture is measured to be approximately 6.5, implying that the $SrTiO_3$ particle surface is positively charged. On the other hand, the oxygen-containing functional groups of graphene oxide (such as carboxylic acid -COOH and hydroxyl -OH) are deprotonated when it immersed in water, which leads to negative charges created on graphene oxide [27]. As a result, the $SrTiO_3$ particles are expected to be adsorbed onto the graphene oxide sheets through electrostatic interactions. Upon UV-light irradiation, electrons and holes are produced on the conduction band (CB) and valence band (VB) of the $SrTiO_3$ particles, respectively. The photogenerated holes are captured by ammonium oxalate that is a hole scavenger [28], leaving behind the photogenerated electrons on the surface of the $SrTiO_3$ particles. The electrons are injected from the $SrTiO_3$ particles into the graphene oxide and react with its oxygen-containing functional groups to reduce graphene oxide.

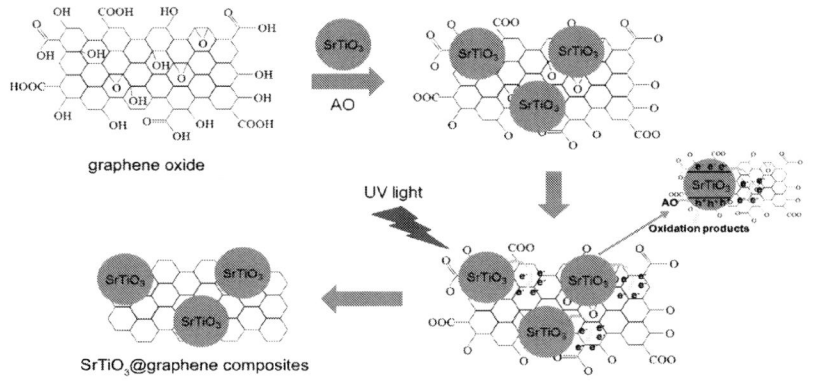

Figure 1: Schematic illustration of the photocatalytic reduction process of graphene oxide by UV light-irradiated $SrTiO_3$ nanoparticles.

Figure 2 shows the FTIR spectra of graphene oxide, $SrTiO_3$ particles, and $SrTiO_3$-graphene (10%) composites. In the spectrum of graphene oxide, the absorption peak at 1,726 cm^{-1} is caused by the C=O stretching vibration of the COOH group. The peak at 1,620 cm^{-1} is attributed to the C=C skeletal vibration of the graphene sheets. The absorption peak of O-H deformation vibrations in C-OH can be seen at 1,396 cm^{-1}. The absorption bands at around 1,224 and 1,050 cm^{-1} are assigned to the C-O stretching vibration. For the $SrTiO_3$ particles, the broad absorption bands at around 447 and 625 cm^{-1} correspond to TiO$_6$ octahedron bending and stretching vibration, respectively [29]. The absorption peak at around 1,630 cm^{-1} is due to the bending vibration of H-O-H from the adsorbed H$_2$O. In the spectrum of the SrTiO$_3$-graphene composites, the characteristic peaks of SrTiO$_3$ are detected. The absorption peak at 1,630 cm^{-1} is the overlay of the vibration peak of H-O-H from H$_2$O and C=C skeletal vibration peak in the graphene sheets. However, the absorption peaks of oxygen-containing functional groups, being characteristic for graphene oxide, disappear. The results demonstrate that graphene oxide is completely reduced to graphene during the photocatalytic reduction process.

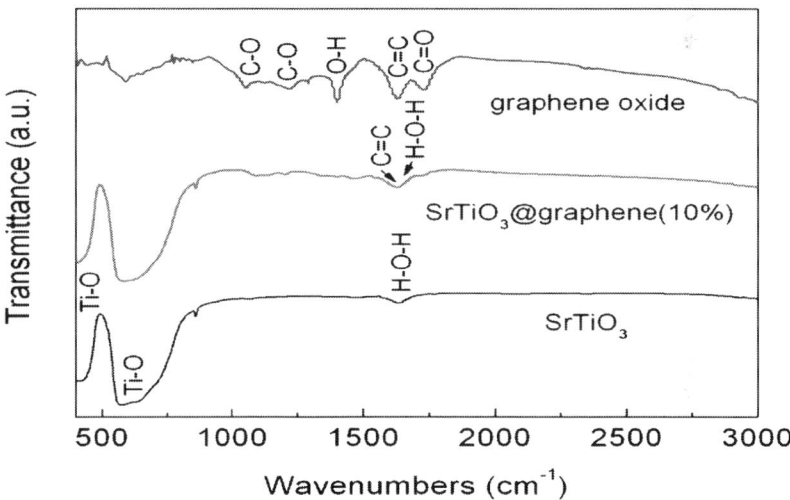

Figure 2: FTIR spectra of graphene oxide, SrTiO$_3$ particles, and SrTiO$_3$-graphene (10%) composites.

Figure 3 shows the XRD patterns of the SrTiO$_3$ particles and the SrTiO$_3$-graphene (10%) composites. It is seen that all the diffraction peaks for the bare SrTiO$_3$ particles and the composites can be index to the cubic structure of SrTiO$_3$, and no traces of impurity phases are detected. This indicates that the SrTiO$_3$ particles undergo no structural change after the photocatalytic reduction of graphene oxide. In addition, no apparent diffraction peaks of graphene in the composites are observed, which is due to the low content and relatively weak diffraction intensity of the graphene.

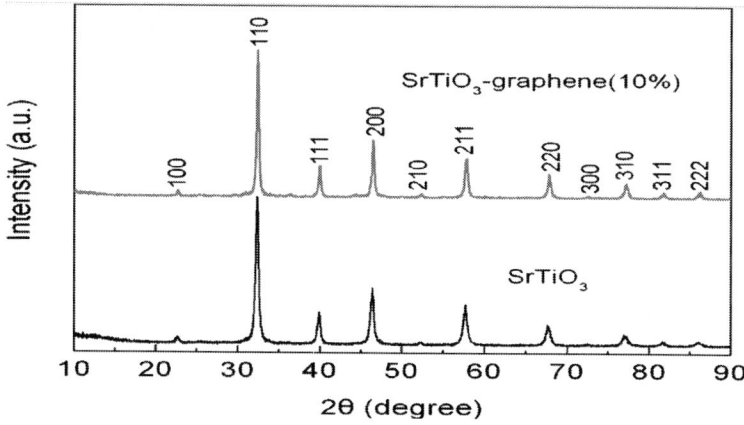

Figure 3: XRD patterns of the SrTiO$_3$ particles and SrTiO$_3$-graphene (10%) composites.

Figure 4a shows the TEM image of graphene oxide, indicating that it has a typical two-dimensional sheet structure with crumpled feature. Figure 4b shows the TEM image of the SrTiO$_3$ particles, revealing that the particles are nearly spherical in shape with an average size of about 55 nm. The TEM image of the SrTiO$_3$-graphene(10%) composites is presented in Figure 4c, from which one can see that the SrTiO$_3$ particles are well assembled onto the graphene sheet.

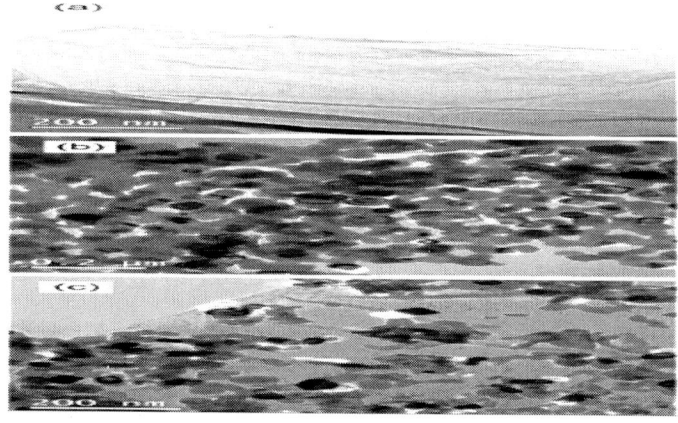

Figure 4: TEM images of (a) graphene oxide, (b) SrTiO$_3$ particles, and (c) SrTiO$_3$-graphene (10%) composites.

Figure 5a shows the UV-visible diffuse reflectance spectra of the SrTiO$_3$ particles and SrTiO$_3$-graphene composites. The composites display continuously enhanced light absorbance over the whole wavelength range with increasing graphene content. This can be attributed to the strong light absorption of graphene in the UV-visible light region [30]. Figure 5b shows the corresponding first derivative of the reflectance (R) with respect to wavelength   (i.e., dR/d), where the peak wavelength is characterized to be the absorption edge of the samples. It is seen that the SrTiO$_3$particles and composites present two absorption peaks in the derivative spectra. The strong and sharp absorption edge at approximately 370 nm is suggested to be attributed to the electron transition from valence band to conduction band. In comparison to the SrTiO$_3$ particles, the SrTiO$_3$-graphene composites show almost no shift in this absorption edge, indicating that the effect of graphene on the band structure of SrTiO$_3$ can be neglected. From this absorption edge, the E_g of the samples is obtained to be approximately 3.35 eV. In addition, the relatively weak absorption edge at approximately 335 nm may be ascribed to the surface effects.

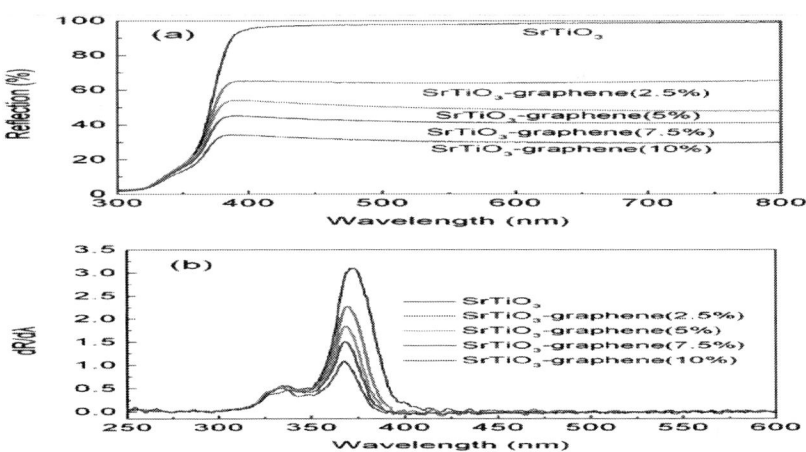

Figure 5: Diffuse reflectance spectra and corresponding first derivative. (a) Diffuse reflectance spectra of the samples. (b)Corresponding first derivative of diffuse reflectance spectra.

The photocatalytic activity of the SrTiO$_3$-graphene composites was evaluated by the degradation of AO7 under UV light irradiation. Figure 6 shows the photocatalytic degradation of AO7 over the SrTiO$_3$-

graphene composites as a function of irradiation time (*t*). The blank experiment result is also shown in Figure 6, from which one can see that AO7 is hardly degraded under UV light irradiation without photocatalysts, and its degradation percentage is less than 8% after 6 h of exposure. After the 6-h irradiation in the presence of SrTiO$_3$ particles, about 51% of AO7 is observed to be degraded. When the SrTiO$_3$ particles assembled on the graphene sheets, the obtained samples exhibit higher photocatalytic activity than the bare SrTiO$_3$ particles. In these composites, the photocatalytic activity increases gradually with increasing graphene content and achieves the highest value when the content of graphene reaches 7.5%, where the degradation of AO7 is about 88% after irradiation for 6 h. Further increase in graphene content leads to the decrease of the photocatalytic activity.

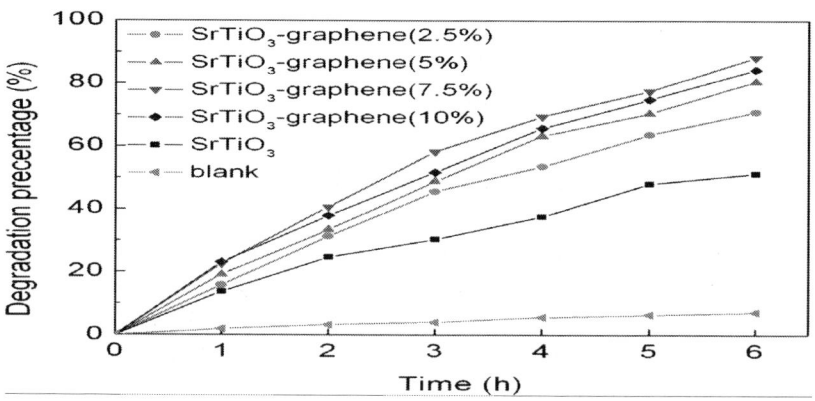

Figure 6: Photocatalytic degradation of AO7 over SrTiO$_3$ particles and SrTiO$_3$-graphene composites. This degradation is a function of irradiation time, along with the blank experiment result.

Figure 7 shows the PL spectra of the TA solution after reacting for 6 h over the UV light-irradiated SrTiO$_3$ particles and SrTiO$_3$-graphene(7.5%) composites. The blank experiment result indicates almost no PL signal at 429 nm after irradiation without photocatalyst. On irradiation in the presence of the SrTiO$_3$ particles, the PL signal centered around 429 nm is obviously detected, revealing the generation of ·OH radicals. When the SrTiO$_3$-graphene composites are used as the photocatalyst, the PL signal becomes more intense, suggesting that the yield of the ·OH radicals is enhanced over the irradiated composites.

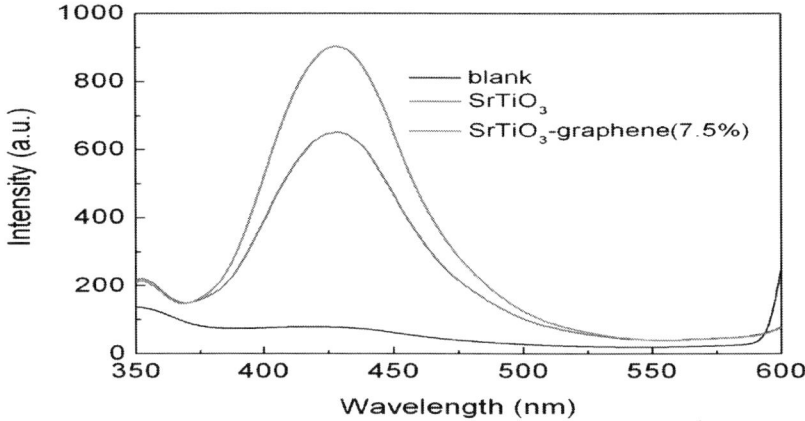

Figure 7: PL spectra of the TA solution after reacting for 6 h over the irradiated samples. The blank experiment result is also shown.

Generally, h^+, $\cdot OH$, $\cdot O_2$, and H_2O_2 are thought to be the main active species responsible for the dye degradation [31]. It is known that ethanol is a scavenger for $\cdot OH$, and KI is a scavenger for both $\cdot OH$ and h^+ [32,33]. By investigating the effect of ethanol and KI on the photocatalytic efficiency of the composites toward the AO7 degradation, we can clarify the role of h^+ and $\cdot OH$ in the photocatalysis. The role of $\cdot O_2$ and H_2O_2, which are derived from the reaction between dissolved O_2 and photogenerated e^-, on the dye degradation can be examined by investigating the effect of N_2 on the photocatalytic efficiency since the dissolved O_2 can be removed from the solution by the N_2-purging procedure. Figure 8 shows the effect of N_2 (bubbled at a rate of 0.1 L min⁻¹), ethanol (10% by volume), and KI (2×10^{-3} mol L⁻¹) on the degradation percentage of AO7 after 6 h of photocatalysis. It is demonstrated that when adding ethanol to the reaction solution, the photocatalytic degradation of AO7 undergoes a substantial decrease, from approximately 88% under normal condition to approximately 40% on addition of ethanol. This suggests that $\cdot OH$ radical is an important active species responsible for the dye degradation. Figure 7 provides direct evidence showing the generation of $\cdot OH$ radicals over the irradiated SrTiO₃-graphene composites. The addition of KI to the reaction solution results in a higher suppression of the photocatalytic efficiency compared to the addition of ethanol, where only 16% of AO7 is caused to be degraded, indicating that the photogenerated h^+ also plays a role in the degradation of AO7. In addition, the photocatalytic efficiency decreases slightly under N_2-purging condition, implying comparatively minor role of $\cdot O_2$ and/or H_2O_2 for the dye degradation.

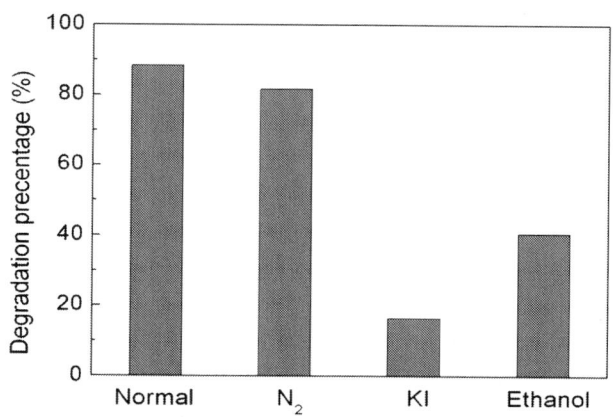

Figure 8: Effects of N_2, ethanol, and KI on the degradation percentage of AO7 over $SrTiO_3$-graphene (7.5%) composites. The irradiation time is 6 h.

In order to understand the photocatalytic mechanism of semiconductor-based photocatalysts, it is essential to determine their energy-band potentials since the redox ability of photogenerated carriers is associated with energy-band potentials of photocatalysts. The conduction band and valence band potentials of $SrTiO_3$ can be calculated using the following relation [34]:

$$E_{CB} = X - E^e - 0.5E_g$$
$$E_{VB} = E_{CB} + E_g,$$

(1)

where X is the absolute electronegativity of $SrTiO_3$ (defined as the arithmetic mean of the electron affinity and the first ionization of the constituent atoms) and estimated to be 5.34 eV according to the data reported in the literature [35, 36], E^e is the energy of free electrons on the hydrogen scale (4.5 eV), and E_g is the bandgap energy of $SrTiO_3$ (3.35 eV). The conduction band and valence band potentials of $SrTiO_3$ vs. normal hydrogen electrode (NHE) are therefore calculated to be $E_{CB} = -0.84$ V and $E_{VB} = +2.51$ V, respectively.

Based on the obtained experimental results, a possible photocatalytic mechanism of $SrTiO_3$-graphene composites toward the degradation of AO7 is schematically shown in Figure 9. When $SrTiO_3$ is irradiated with light of energy greater than its bandgap energy, electrons are excited to the conduction band from the valence band, thus creating electron–hole pairs (Equation 2). Generally, most of the photogenerated electrons and holes recombine rapidly, and only a few of them participate in redox reactions. It is noted that graphene, which is an excellent electron acceptor and conductor, has a Fermi level (-0.08 V vs. NHE [37]) positive to the conduction band potential of $SrTiO_3$ (-0.84 V). When $SrTiO_3$ particles are assembled onto graphene sheets, the photogenerated electrons can readily transfer from the conduction band of $SrTiO_3$ to graphene (Equation 3). Thus, the recombination of electron–hole pairs can be effectively suppressed in the composites, which leads to an increased availability of electrons and holes for the photocatalytic reactions. The Fermi level of graphene is positive to the redox potential of $O_2/\cdot O_2$ (-0.13 V vs. NHE) but negative to that of O_2/H_2O_2 (+0.695 vs. NHE) [31, 38]. This implies that the photogenerated e^- which transferred onto the graphene cannot thermodynamically react with O_2 to produce $\cdot O_2$, but can react with O_2 and H^+ to produce H_2O_2 (Equation 4). H_2O_2 is an active species that can cause dye degradation, and moreover, H_2O_2 can also participate in the reactions as described in Equations 5 and 6 to form another active species $\cdot OH$. The valence band potential of $SrTiO_3$ (+2.51 V) is positive to the redox potential of $OH^-/\cdot OH$ (+1.89 V vs. NHE) [39], indicating that the photogenerated h^+ can react with OH^- to produce $\cdot OH$ (Equation 7). As a consequence, the active species $\cdot OH$, h^+, and H_2O_2 work together to degrade AO7 (Equation 8).

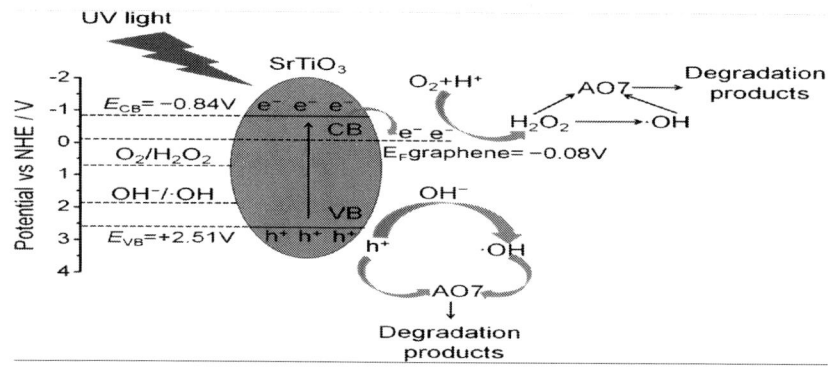

Figure 9: Schematic illustration of the photocatalytic mechanism of SrTiO₃-graphene composites toward the degradation of AO7.

$$SrTiO_3 + h\nu \rightarrow SrTiO_3\left(e^- + h^+\right)$$

(2

$$e^- + Graphene \rightarrow Graphene\ (e^-)$$

(3)

$$2\ Graphene\ (e^-) + O_2 + 2H^+ \rightarrow 2\ Graphene + H_2O_2$$

(4)

$$H_2O_2 + h\nu \rightarrow 2 \cdot OH$$

(5)

$$H_2O_2 + e^- \rightarrow \cdot OH + OH^-$$

(6)

$$h^+ + OH^- \rightarrow \cdot OH$$

(7)

$$h^+,\ \cdot OH,\ or\ H_2O_2 + AO7 \rightarrow Degradation\ products$$

(8)

From Figure 6, it is found that the photocatalytic activity of the composites is highly related to the content of graphene, which can be explained as follows. With raising the graphene content, the amount of SrTiO₃ particles decorated on the surface of graphene is expected to increase, thus providing more photogenerated carriers for the photocatalytic reaction. When the graphene content in the composites reaches 7.5%, the SrTiO$_3$ particles are decorated sufficiently, consequently leading to the achievement of the highest photocatalytic activity. However, with further increasing graphene content above 7.5%, the photocatalytic efficiency begins to exhibit a decreasing trend. The possible reason is that the excessive graphene may shield the light and decrease the photon absorption by the SrTiO$_3$ particles, and moreover, the amount of available surface active sites tends to be reduced due to an increasing coverage of graphene onto the surface of the SrTiO$_3$ particles.

Besides the photocatalytic activity, the reusability of photocatalysts is another crucial factor for their practical applications. The stability of the SrTiO$_3$-graphene (7.5%) composites is examined by the recycling photocatalytic experiment, as shown in Figure 10. It reveals that the degradation percentage of AO7 maintains 80% to 88% for five consecutive recycles. The tiny or negligible lose of the photocatalytic efficiency indicates the excellent photocatalytic reusability of the as-prepared SrTiO$_3$-graphene composites. Figure 11 shows the XRD patterns of the composites before and after the recycle experiment, revealing no obvious crystal structure changes. Figure 12 shows the TEM images of the composites before and after the recycle experiment, from which one can see that SrTiO$_3$ particles are still well decorated on the graphene sheets.

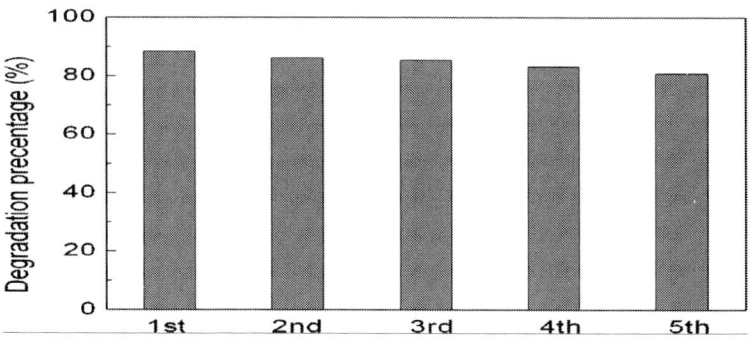

Figure 10: Degradation percentage of AO7 after irradiation for 6 h over Sr-TiO$_3$-graphene (7.5%) composites during the five photocatalytic cycles.

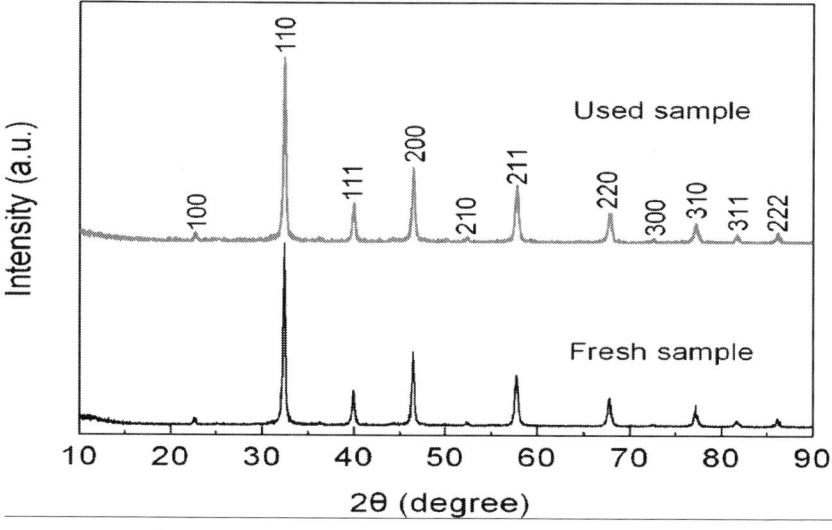

Figure 11: XRD patterns of SrTiO$_3$-graphene (7.5%) composites before and after the photocatalytic experiment.

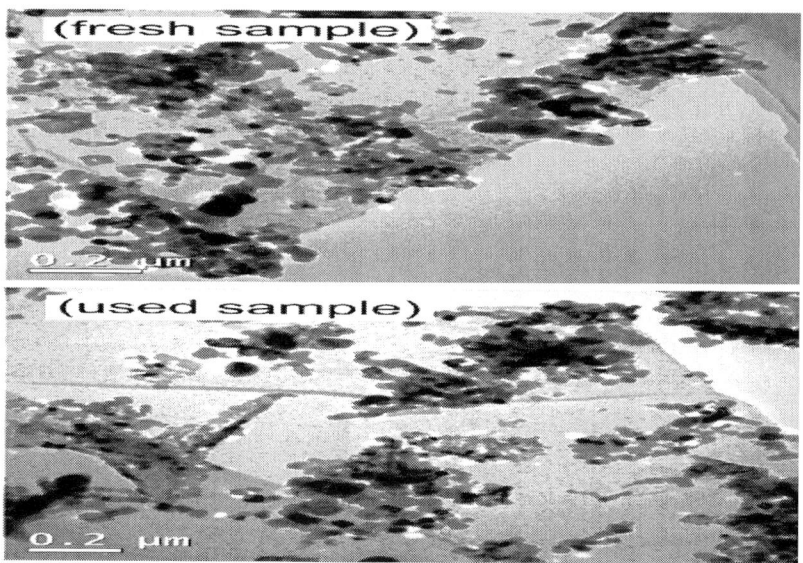

Figure 12: TEM images of the SrTiO$_3$-graphene (7.5%) composites before (top) and after (bottom) the photocatalytic experiment.

CONCLUSIONS

$SrTiO_3$-graphene nanocomposites were prepared by irradiating the mixture solution of $SrTiO_3$ nanoparticles and graphene oxide sheets, during which graphene oxide receives electrons from the excited $SrTiO_3$ nanoparticles to be reduced to graphene, simultaneously leading to the assembly of $SrTiO_3$ nanoparticles onto graphene sheets. Compared to the bare $SrTiO_3$ nanoparticles, the as-prepared $SrTiO_3$-graphene composites exhibit an enhanced photocatalytic activity for the degradation of AO7 under irradiation of UV light. This can be attributed to the effective separation of photogenerated electron–hole pairs due to the electron transfer from $SrTiO_3$ to graphene and, hence, increased availability of electrons and holes for the photocatalytic reaction. The enhanced generation of $\cdot OH$ radicals is observed over the irradiated $SrTiO_3$-graphene composites compared to the bare $SrTiO_3$ nanoparticles. The photocatalytic efficiency is slightly deceased by purging with N_2 but is significantly suppressed by the addition of ethanol and KI (especially for the latter). Based on the experimental results, $\cdot OH$, h^+, and H_2O_2 are suggested to be the main active species causing the dye degradation.

AUTHORS' CONTRIBUTIONS

HY and TX conceived the idea of experiments. TX, LD, JM, and HZ carried out the preparation and characterization of the samples. HY, TX, and JD analyzed and discussed the results of the experiments. TX drafted the manuscript. HY improved the manuscript. All authors read and approved the final manuscript.

ACKNOWLEDGEMENTS

This work was supported by the National Natural Science Foundation of China (Grant No. 51262018) and the Hongliu Outstanding Talents Foundation of Lanzhou University of Technology (Grant No. J201205).

REFERENCES

1.   Mills A, Davies RH, Worsley D: Water purification by semiconductor photocatalysis. *Chem Soc Rev* 1993, 22:417-425.

2.   Hofmann MR, Martin ST, Choi W, Bahnemann DW: Environmental applications of semiconductor photocatalysis. *Chem Rev* 1995, 95:69-96.

3.   Zheng Z, Huang B, Qin X, Zhang X, Dai Y: Facile synthesis of $SrTiO_3$ hollow microspheres built as assembly of nanocubes and their associated photocatalytic activity. *J Colloid Interface Sci* 2011, 358:68-72.

4.   Kato H, Kobayashi M, Hara M, Kakihana M: Fabrication of $SrTiO_3$ exposing characteristic facets using molten salt flux and improvement of photocatalytic activity for water splitting. *Catal Sci Technol* 2013, 3:1733-1738.

5.   da Silva LF, Avansi W, Andres J, Ribeiro C, Moreira ML, Longo E, Mastelaro VR: Long-range and short-range structures of cube-like shape $SrTiO_3$ powders: microwave-assisted hydrothermal synthesis and photocatalytic activity. *Phys Chem Chem Phys* 2013, 15:12386-12393.

6.   Kuang Q, Yang S: Template synthesis of single-crystal-like porous $SrTiO_3$ nanocube assemblies and their enhanced photocatalytic hydrogen evolution. *ACS Appl Mat Interfaces* 2013, 5:3683-3690.

7.   Cao T, Li Y, Wang C, Shao C, Liu Y: A facile in situ hydrothermal method to $SrTiO_3/TiO_2$ nanofiber heterostructures with high photocatalytic activity. *Langmuir* 2011, 27:2946-2952.

8.   Puangpetch T, Chavadej S, Sreethawong T: Hydrogen production over Au-loaded mesoporous-assembled $SrTiO_3$ nanocrystal photocatalyst: effects of molecular structure and chemical properties of hole scavengers. *Energy Convers Manage* 2011, 52:2256-2261.

9.   Guoa J, Ouyang S, Li P, Zhang Y, Kako T, Ye J: A new heterojunction $Ag_3PO_4/Cr-SrTiO_3$ photocatalyst towards efficient elimination of gaseous organic pollutants under visible light irradiation. *Appl Catal B Environ* 2013, 134–135:286-292.

10. Zou F, Jiang Z, Qin X, Zhao Y, Jiang L, Zhi J, Xiao T, Edwards PP: Template-free synthesis of mesoporous N-doped $SrTiO_3$ perovskite with high visible-light-driven photocatalytic activity. *Chem Commun* 2012, 48:8514-8516.

11. Bolotin KI, Sikes KJ, Jiang Z, Klima M, Fudenberg G, Hone J, Kim P, Stormer HL: Ultrahigh electron mobility in suspended graphene. *Solid State Commun* 2008, 146:351-355.

12. Balandin AA, Ghosh S, Bao W, Calizo I, Teweldebrhan D, Miao F, Lau CN: Superior thermal conductivity of single-layer graphene. *Nano Lett* 2008, 8:902-907.

13. Frank IW, Tanenbaum DM, Van Der Zande AM, McEuen PL: Mechanical properties of suspended graphene sheets. *J Vac Sci Technol B* 2007, 25:2558-2561.

14. Xu TG, Zhang LW, Cheng HY, Zhu YF: Significantly enhanced photocatalytic performance of ZnO via graphene hybridization and the mechanism study. *Appl Catal B Environ* 2011, 101:382-387.

15. Cuong TV, Pham VH, Tran QT, Chung JS, Shin EW, Kim JS, Kim EJ: Optoelectronic properties of graphene thin films prepared by thermal reduction of graphene oxide. *Mater Lett* 2010, 64:765-767.

16. Lü W, Chen J, Wu Y, Duan L, Yang Y, Ge X: Graphene-enhanced visible-light photocatalysis of CdS particles for wastewater treatment. *Nanoscale Res Lett* 2014, 9:148.

17. Gao M, Peh CKN, Ong WL, Ho GW: Green chemistry synthesis of a nanocomposite graphene hydrogel with three-dimensional nanomesopores for photocatalytic H_2production. *RSC Advances* 2013, 3:13169-13177.

18. Liu X, Pan L, Zhao Q, Lv T, Zhu G, Chen T, Lu T, Sun Z, Sun C: UV-assisted photocatalytic synthesis of ZnO-reduced graphene oxide composites with enhanced photocatalytic activity in reduction of Cr(VI). *Chem Eng J* 2012, 183:238-243.

19. Wong TJ, Lim FJ, Gao M, Lee GH, Ho GW: Photocatalytic H_2 production of composite one-dimensional TiO_2 nanostructures of different morphological structures and crystal phases with graphene. *Catal Sci Technol* 2013, 3:1086-1093.

20. Bell NJ, Ng YH, Du A, Coster H, Smith SC, Amal R: Understanding the enhancement in photoelectrochemical properties of photocatalytically prepared TiO_2-reduced graphene oxide composite. *J Phys Chem C* 2011, 115:6004-6009.

21. Akhavan O: Graphene nanomesh by ZnO nanorod photocatalysts. *ACS Nano* 2010, 7:4174-4780.

22. Li Z, Zhou Z, Yun G, Shi K, Lv X, Yang B: High-performance solid-state supercapacitors based on graphene-ZnO hybrid nanocomposites. *Nanoscale Res Lett* 2013, 8:473.

23. Yan Z, Ma L, Zhu Y, Lahiri I, Hahm MG, Liu Z, Yang S, Xiang C, Lu W, Peng Z, Sun Z, Kittrell C, Lou J, Choi W, Ajayan PM, Tour JM: Three-dimensional metal-graphene-nanotube multifunctional hybrid materials. *ACS Nano* 2013, 7:58-64.

24. Liang Y, Li Y, Wang H, Zhou J, Wang J, Regier T, Dai H: Co_3O_4 nanocrystals on graphene as a synergistic catalyst for oxygen reduction reaction. *Nat Mater* 2011, 10:780-786.

25. Xian T, Yang H, Dai JF, Wei ZQ, Ma JY, Feng WJ: Photocatalytic properties of $SrTiO_3$ nanoparticles prepared by a polyacrylamide gel route. *Mater Lett* 2011, 21–22:3254-3257.

26. Kosmulski M: pH-dependent surface charging and points of zero charge. IV. Update and new approach. *J Colloid Interface Sci* 2009, 337:439-448.

27. Talyzin AV, Hausmaninger T, You S, Szabob T: The structure of graphene oxide membranes in liquid water, ethanol and water-ethanol mixtures. *Nanoscale* 2014, 6:272-281.

28. Liu W, Wang M, Xu C, Chen S, Fu X: Significantly enhanced visible-light photocatalytic activity of g-C_3N_4 via ZnO modification and the mechanism study. *J Mol Catal A Chem* 2013, 9–15:368-369.

29. Last JT: Infrared-absorption studies on barium titanate and related materials. *Phys Rev* 1957, 105:1740-1750.

30. Zhao D, Sheng G, Chen C, Wang X: Enhanced photocatalytic degradation of methylene blue under visible irradiation on graphene@TiO_2 dyade structure. *Appl Catal B Environ* 2012, 111–112:303-308.

31. Teoh WY, Scott JA, Amal R: Progress in heterogeneous photocatalysis: from classical radical chemistry to engineering

nanomaterials and solar reactors. *J Phys Chem Lett* 2012, 3:629-639.

32. Daneshvar N, Salari D, Khataee AR: Photocatalytic degradation of azo dye acid red 14 in water: investigation of the effect of operational parameters. *J Photochem Photobiol A Chem* 2003, 157:111-116.

33. Li YY, Wang JS, Yao HC, Dang LY, Li Z: Efficient decomposition of organic compounds and reaction mechanism with BiOI photocatalyst under visible light irradiation. *J Mol Catal A Chem* 2011, 334:116-122.

34. Morrison SR: *Electrochemistry at Semiconductor and Oxidized Metal Electrode*. New York: Plenum; 1980.

35. Hotop H, Lineberger WC: Binding energies in atomic negative ions. *J Phys Chem Ref Data* 1975, 4:539-576.

36. Andersen T, Haugen HK, Hotop H: Binding energies in atomic negative ions: III. *J Phys Chem Ref Data* 1999, 28:1511-1533.

37. Zhang J, Yu J, Jaroniec M, Gong JR: Noble metal-free reduced graphene oxide-$Zn_xCd_{1-x}S$ nanocomposite with enhanced solar photocatalytic H_2-production performance. *Nano Lett* 2012, 12:4584-4589.

38. Arai T, Yanagida M, Konishi Y, Iwasaki Y, Sugihara H, Sayama K: Efficient complete oxidation of acetaldehyde into CO_2 over $CuBi_2O_4/WO_3$ composite photocatalyst under visible and UV light irradiation. *J Phys Chem C* 2007, 111C:7574-7577.

39. Tachikawa T, Fujitsuka M, Majima T: Mechanistic insight into the TiO_2 photocatalytic reactions: design of new photocatalysts. *J Phys Chem C* 2007, 111C:5259-5275.

Citations

CHAPTER 1

Gui-Fang Huang, Zhi-Li Ma, Wei-Qing Huang, et al., "Ag_3PO_4 Semiconductor Photocatalyst: Possibilities and Challenges," Journal of Nanomaterials, vol. 2013, Article ID 371356, 8 pages, 2013. doi:10.1155/2013/371356.

CHAPTER 2

Sandra Andrea Mayén-Hernández, David Santos-Cruz, Francisco de Moure-Flores, et al., "Optical, Electrical and Photocatalytic Properties of the Ternary Semiconductors $Zn_xCd_{1-x}S$, $Cu_xCd_{1-x}S$ and $Cu_xZn_{1-x}S$,"International Journal of Photoenergy, vol. 2014, Article ID 158782, 8 pages, 2014. doi:10.1155/2014/158782.

CHAPTER 3

Nurhidayatullaili Muhd Julkapli, Samira Bagheri, and Sharifah Bee Abd Hamid, "Recent Advances in Heterogeneous Photocatalytic Decolorization of Synthetic Dyes," The Scientific World Journal, vol. 2014, Article ID 692307, 25 pages, 2014. doi:10.1155/2014/692307.

CHAPTER 4

Dina Mamdouh Fouad and Mona Bakr Mohamed, "Comparative Study of the Photocatalytic Activity of Semiconductor Nanostructures and Their Hybrid Metal Nanocomposites on the Photodegradation of Malathion," Journal of Nanomaterials, vol. 2012, Article ID 524123, 8 pages, 2012, doi:10.1155/2012/524123.

CHAPTER 5

Hongchao Ma, Xiaohong Cheng, Chun Ma, et al., "Synthesis, Characterization, and Photocatalytic Activity of N-Doped ZnO/ZnS Composites," International Journal of Photoenergy, vol. 2013, Article ID 625024, 8 pages, 2013. doi:10.1155/2013/625024.

CHAPTER 6

Dipak Nipane, S. R. Thakare, and N. T. Khati, "Synthesis of Novel ZnO Having Cauliflower Morphology for Photocatalytic Degradation Study," Journal of Catalysts, vol. 2013, Article ID 940345, 8 pages, 2013, doi:10.1155/2013/940345.

CHAPTER 7

Sulaiman N Basahel, Tarek T Ali, Mohamed Mokhtarm, and Katabathini Narasimharao, Influence of Crystal Structure of Nanosized ZrO_2 on Photocatalytic Degradation of Methyl Orange, doi: 10.1186/s11671-015-0780-z.

CHAPTER 8

Xiangyun Zeng, Jiao Yang, Liuxue Shi, Linjie Li, and Meizhen Gao, Synthesis of Multi-shelled ZnO Hollow Microspheres and their Improved Photocatalytic Activity, doi:10.1186/1556-276X-9-468.

CHAPTER 9

Tao Xian, Hua Yang, Lijing Di, Jinyuan Ma, Haimin Zhang, and Jianfeng Dai, Photocatalytic Reduction Synthesis of $Srtio_3$-graphene Nanocomposites and Their Enhanced Photocatalytic Activity, doi:10.1186/1556-276X-9-327.

Index